Elements
of
Astronomy

Elements
of
Astronomy

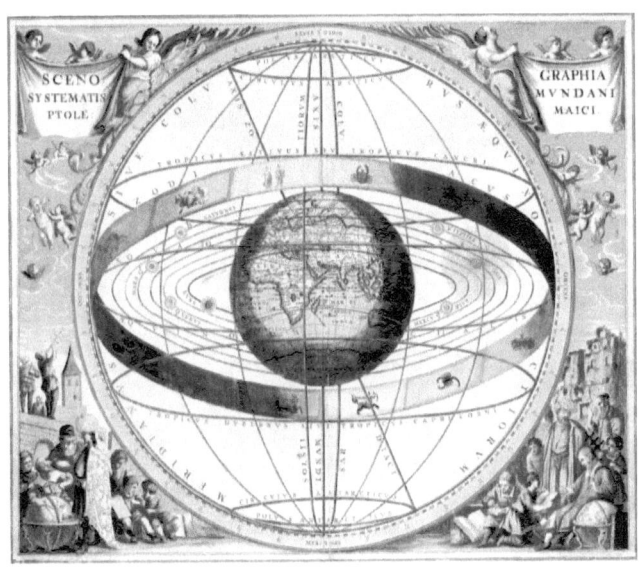

BY
Simon Newcomb, PH.D., LL.D.

Athens ‡ Manchester

Elements of Astronomy

Old Book Publishing Ltd

Book Cover Design: Old Book Publishing Ltd

Copyright © 2011 Old Book Publishing Ltd
All rights reserved.

Title of original: Elements of Astronomy
Originally published in 1890

Copyright © 1900 Simon Newcomb

Cover Image: From Andreas Cellarius Harmonia Macrocosmica, 1660/61. Chart showing signs of the zodiac and the solar system with world at centre.

ISBN-10: 1-78107-048-2
ISBN-13: 978-1-78107-048-2

EDITOR'S NOTE

Old Book Publishing Ltd takes care in preserving the wording and images of the original books. For this reason we have invested in technology that enables us to enhance the quality of such reproduction. This investment helps overcome problems encountered when reproducing old books, such as stains, coloured paper, discolouration of ink, yellowed pages, see-through and onion skin type paper.

This reproduction book, produced from digital images of the original, may contain occasional defects such as missing pages or blemishes due to the original source content or were introduced by the scanning process.

These are scanned pages and the quality of print represents accurately the print quality of the original book, though we may have been able to enhance it.

As this book has been scanned and/or reformatted from the original we cannot guarantee that it is error-free or contains the full content of the original.

However, we believe that this work is culturally important, and despite its imperfections, have elected to bring it back into print as part of our commitment to the preservation of printed works.

Old Book Publishing

Frontispiece TELESCOPIC VIEWS OF THE PLANETS.

ELEMENTS OF ASTRONOMY

BY

SIMON NEWCOMB, Ph.D., LL.D.

FORMERLY PROFESSOR OF MATHEMATICS AND ASTRONOMY
JOHNS HOPKINS UNIVERSITY

NEW YORK ∴ CINCINNATI ∴ CHICAGO
AMERICAN BOOK COMPANY

COPYRIGHT, 1900, BY
SIMON NEWCOMB.

EL. OF ASTRON.

PREFACE

Two objects have been kept in view in preparing this little book. One was to condense those facts and laws of the science which are of most interest and importance to the general intelligent public within so small a compass as not to make a very serious addition to the curriculum of the high school or college. The other was so to present the subject that as little formal mathematics as possible should be necessary to its mastery.

Of the first object little need be said. The typical person constantly kept in mind has been the inquiring layman seeking to know something of the heavenly bodies and their relation to the earth, including such subjects of human interest as the changing seasons, the measure of time, and the varying aspects of the planets.

The second object involves more serious questions. Can an idea of the laws and phenomena of the celestial motions be conveyed to a pupil who has not completed the regular course in geometry and physics? The author believes that it can. It cannot, indeed, be denied that the professional astronomer, engineer, surveyor, and navigator who are to make astronomical observations and computations must have a fairly complete training in at least the elementary branches of mathematics. But this training is not essential to him who desires only a command of general ideas, without proposing to make technical applications of the science. What is really essential are those conceptions of motion and form which one may derive from everyday observation, and the understanding of a few elementary definitions in geometry and physics. Our modern system of education wisely endeavors to implant such conceptions and to teach the corresponding definitions at an earlier age than that when the growing youth is expected to commence a course of formal mathematics.

The author hopes that the early chapters are the only ones that will offer any difficulty to an intelligent pupil prepared for a high school course. Here it is believed that every difficulty may be overcome by two very simple measures on the part of the teacher. One is to point out, approximately, the actual position of the celestial poles and equator and the apparent diurnal courses of the sun and stars, as they might be seen in the mind's eye from the schoolroom or the field. The object of this is that the learner may conceive the phenomena he is studying as if seen in the sky. The other is to see that the learner correctly apprehends the meaning of the figures representing points, circles, and motions on the celestial sphere; especially, that he always imagines himself looking at the objects represented as if he were at the center of the sphere.

For this last suggestion and for other valuable hints, the author takes much pleasure in acknowledging his indebtedness to Mr. Edward P. Jackson, teacher of Physics in the Boston Latin School.

CONTENTS

CHAPTER		PAGE
I.	RELATION OF THE EARTH TO THE HEAVENS	9

    1. Introduction. 2. Ideas of Motion. 3. The Earth. 4. The Celestial Sphere. 5. Perspective of Plane and Line. 6. Angular Measure on the Celestial Sphere. 7. The Relation of the Horizon to the Celestial Sphere. 8. The Diurnal Motion. 9. Celestial Equator and Poles. 10. The Meridian. 11. Diurnal Motion in Different Latitudes. 12. Right Ascension and Declination. 13. Correspondence of the Terrestrial and Celestial Spheres.

II.	THE REVOLUTION OF THE EARTH ROUND THE SUN . .	32

    1. The Earth as a Planet. 2. Annual Motion of the Earth round the Sun. 3. How the Sun shines on the Earth at Different Seasons. 4. Apparent Motion of the Sun — The Zodiac. 5. Seasons in the Two Hemispheres. 6. The Solar and Sidereal Years. 7. Precession of the Equinoxes.

III.	OF TIME	48

    1. Diurnal Motion of the Sun and Stars. 2. Mean and Apparent Time; Inequality of Apparent Time. 3. Local Time and Longitude. 4. Standard Time.

IV.	OBSERVATION AND MEASUREMENT OF THE HEAVENS . .	56

    1. Refraction of Light. 2. Lenses and Object Glasses. 3. The Refracting Telescope. 4. The Equatorial Telescope. 5. The Reflecting Telescope. 6. Great Telescopes. 7. Meridian Instruments. 8. The Spectroscope and its Use. 9. Semidiameter and Parallax. 10. The Aberration of Light.

V.	GRAVITATION	80

    1. Force. 2. The Laws of Motion. 3. Universal Gravitation. 4. Weight and Mass. 5. How the Attraction of the Sun keeps the Planets in their Orbits. 6. Centrifugal Force.

VI.	THE EARTH	90

    1. Figure and Magnitude of the Earth. 2. Latitude and Longitude. 3. Length of a Degree. 4. How the Earth is measured. 5. How Latitude and Longitude are determined. 6. Density of the Earth, Gravity, etc. 7. Condition of the Earth's Interior. 8. The Atmosphere. 9. The Zodiacal Light.

VII.	THE SUN	103

    1. Particulars about the Sun. 2. Heat of the Sun. 3. Spots and Rotation of the Sun. 4. Corona and Prominences. 5. Source and Period of the Sun's Heat.

CONTENTS

CHAPTER		PAGE
VIII.	THE MOON AND ECLIPSES	112

1. Distance, Size, and Aspect of the Moon. 2. The Moon's Revolution. 3. The Moon's Phases and Rotation. 4. The Tides. 5. Eclipses of the Moon. 6. The Moon's Orbit and Nodes. 7. Eclipses of the Sun. 8. Recurrence of Eclipses.

IX. THE CALENDAR 133
1. Units of Time. 2. The Julian Calendar. 3. The Gregorian Calendar. 4. The Year. 5. Features of the Church Calendar. 6. The Hours.

X. GENERAL PLAN OF THE SOLAR SYSTEM . . . 140
1. Orbits of the Planets. 2. Kepler's Laws. 3. Structure of the Solar System. 4. Distances of the Planets; Bode's Law. 5. Aspects of the Planets. 6. Apparent Motions of the Planets. 7. Perturbations of the Planets.

XI. THE INNER GROUP OF PLANETS 151
1. The Planet Mercury. 2. The Planet Venus; Aspects of Venus. 3. The Planet Mars; Aspects of Mars. 4. The Minor Planets or Asteroids.

XII. THE FOUR OUTER PLANETS 162
1. The Planet Jupiter. 2. The Satellites of Jupiter. 3. The Planet Saturn. 4. The Rings of Saturn. 5. The Satellites of Saturn. 6. Uranus and its Satellites. 7. Neptune and its Satellite.

XIII. COMETS AND METEORS 176
1. Appearance of a Comet. 2. Comets belong to the Solar System. 3. Orbits of Comets. 4. Remarkable Comets. 5. Constitution of Comets. 6. Meteors. 7. Meteoric Showers.

XIV. THE CONSTELLATIONS 191
1. About the Stars in General. 2. How the Constellations and Stars are named. 3. Description of the Principal Constellations. 4. Constellations Visible in the Evenings of February and March. 5. The Early Summer Constellations. 6. The August Constellations. 7. The November Constellations.

XV. THE STARS AND NEBULÆ 205
1. The Stars are Suns. 2. Proper Motions of the Stars. 3. Motion of the Sun. 4. Motions in the Line of Sight. 5. Distances of the Stars. 6. Variable Stars. 7. Double Stars. 8. Clusters and Nebulæ; Clusters of Stars.

XVI. A BRIEF HISTORY OF ASTRONOMY 225

INDEX 237

ASTRONOMY

CHAPTER I

RELATION OF THE EARTH TO THE HEAVENS

1. Introduction. — When we look at the sky by day we see the sun; by night we see the moon and stars. These, and all other objects which we see in the heavens, are called *heavenly bodies*. *Astronomy* is the science which treats of these bodies.

The heavenly bodies are all of immense size, most of them larger than the earth. They look small because they are so far away. If we could fly from the earth as far as we please, it would look smaller and smaller as we went farther, until at a distance of many millions of miles it would appear as a little star. If we kept on yet farther, it would at last disappear from our sight altogether.

If we lived on one of the heavenly bodies, it would be to us as the earth, and the earth would be seen as a heavenly body.

In trying to think of the relation of the earth to the heavens, we may liken ourselves to microscopic insects living on an apple. To them the apple is a world, than which nothing bigger can be conceived. As this continent is to their apple, so is the universe of stars to our world. We may fancy how their ideas would have to be enlarged to make them comprehend the relations of the Atlantic and Pacific oceans; and then we may try to enlarge ours in the same way to understand the relations of the heavenly bodies.

2. Ideas of Motion. — If we think carefully, we shall see that we can never know that any object is in motion except by comparing its position with that of some other object supposed to be at rest. Inside the cabin of a ship on a smooth sea we are not able to decide whether we are at rest or in motion unless we can look out on the ocean which we suppose to be at rest. Even then water, ship, and everything on the ship, might be carried along by the Gulf Stream without our knowing it. This general fact is expressed by saying that all motion, so far as we can define or know it, is *relative;* that is, it is referred to some object supposed to be at rest.

It follows from this that the motion of an object may be very different according to the body to which it is referred. Suppose, for example, that a man walks from the front to the rear of a railway car running eastward 50 miles an hour. A fellow passenger would say that the man was walking westward at the rate of three miles an hour, because his motion would be referred to the car as if the latter were at rest. But if we refer it to the surface of the earth, he would be going east at the rate of 47 miles an hour. Hence, in speaking of the motion of a body, there must always be some other body, or some position, to which the motion is referred.

In everyday life we commonly refer the motions of things around us to the surface of the earth. In astronomy motions are sometimes referred to the center of the earth, or to the sun, or even to the stars.

3. The Earth. — Some of the following facts are taught in geography, but they are of equal importance in astronomy: —

1. The earth has the form of a spheroid. Its figure is so near that of a globe that the eye could not see any deviation from the spherical form. Hence, we commonly speak of the earth as a globe.

2. We live on the round surface of this globe.

3. Our bodies and everything else on the earth's surface are drawn toward its center by a force called *gravity*. Were it

RELATION OF THE EARTH TO THE HEAVENS

not for gravity, objects on the earth would have no tendency to stay there.

4. It follows that the direction we call *downward* is not the same in any two places, because it is everywhere nearly toward the earth's center. Dwellers on the opposite side of the earth stand with their feet pointing toward us, and are therefore called our *antipodes*.

5. The earth turns continually from west to east on an imaginary line passing through its center, and called its *axis*. The two opposite points in which the axis intersects the surface of the earth are called *poles*. One of these is called the north pole, the other the south pole. The time required to make a revolution is called a *day*.

6. An imaginary circle passing round the earth, equally distant from the two poles, is called the earth's *equator*.

7. The motion of the earth on its axis is so smooth and uniform that we are entirely unconscious of it. Hence it seems to us to be at rest while the heavenly bodies seem to move in the opposite direction, from east toward west.

4. The Celestial Sphere. — When we look up from the earth, the stars seem to be set in a blue vault or dome, which we call the sky. The sky seems to rise high over our heads, and to curve down on every side toward the earth, on which it seems to rest. The sky is not a real object, but only an appearance produced by the blue light reflected from the air to our eyes.

There are as many stars in the heavens by day as by night. The reason we do not see them by day is that our eyes are dazzled by the light of the sky, which is really light reflected by the air. If we could mount above the air, we should see no sky, because there would be no air to reflect the light, and we should see the stars all day as well as all night.

The stars surround us in every possible direction, below our feet as well as above our heads. The earth is in the way of our seeing them when they are below us, but they are then visible to our antipodes.

12 ASTRONOMY

The heavenly bodies are really at very different distances. They appear to us to be at the same distance because our eyes cannot distinguish their distances as less or greater. Hence we fancy them to be on the surface of a hollow sphere, in the

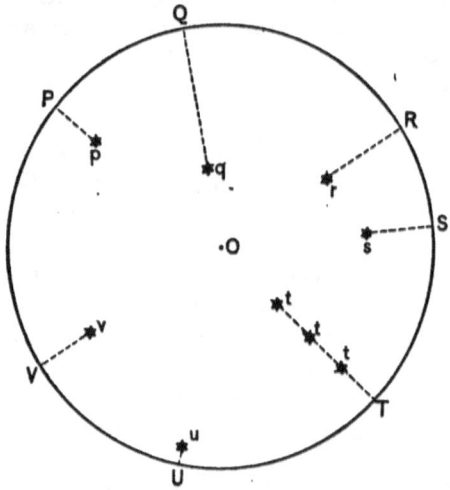

FIG. 1. — Showing how stars *p, q, r, s*, etc., are seen by an observer at *O* as if they all lay on a sphere at the respective points *P, Q, R, S*, etc. The three stars marked *t* are seen as if they were a single star in the position *T*, because, being in the same straight line from the observer, they cannot be distinguished.

center of which we stand. Although this sphere is imaginary, it will help our thoughts to think of it and talk about it as if it were real. It is called the *celestial sphere*.

We must imagine the celestial sphere to be so vast that the earth in its center is a mere point in comparison.

5. Perspective of Plane and Line. — You doubtless know that a plane is a flat surface which may be supposed to extend out all round us as far as we please. When we represent a plane on paper we have to give it a boundary because there is not room enough on the paper to show it extending out without

RELATION OF THE EARTH TO THE HEAVENS 13

limit. We are not to suppose that the circle or other bounding line on the figure is really the boundary of the plane; the latter need not have a boundary.

Let us see how we may represent a plane seen in different ways. Figure 2, *a*, shows a small portion of a plane, bounded by a circle, on which we are looking perpendicularly. This plane coincides with the plane of the paper.

Figure 2, *b*, shows the plane seen obliquely. We must conceive this plane as passing through the plane of the paper.

Figure 2, *c*, shows the same plane seen edgewise. It then looks like a straight line, but must still be conceived as a plane, and as being perpendicular to the plane of the paper.

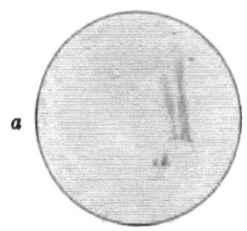

Fig. 2.

6. Angular Measure on the Celestial Sphere.

When we speak of the distance of two heavenly bodies from each other, the word *distance* may have either of two meanings.

The real distance is the length of the line from one body to the other. Hence this is also called *linear distance*.

The *apparent distance* between two heavenly bodies is their distance apart as it appears to us. This is not a line, but the angle between two lines from the observer's eye, one going toward one body and one toward the other. In astronomy we commonly use words expressing distance in this sense; thus we say that the moon and a star are together, or that a star is alongside the moon, when they look so to our eyes, although the star is in reality millions of times farther from us than the moon. Two heavenly bodies appear together when they lie in the same line from the observer.

The apparent distance being an angle, is measured as angles

are measured in other cases, the position of the observer being the vertex of the angle. Imagine a circle on the celestial sphere with the eye of the observer in the center as shown in figure 3.

FIG. 3. — Showing the angle between two stars, *a* and *b*, as seen by an observer. In the figure this angle is about 10°. The figure also shows how degrees are counted round the circle, passing from the line going horizontally to the right. A right angle, or 90°, measures from the horizon *A* to the zenith *B*. Two right angles, or 180°, bring us to the horizon on the left; the third right angle, making 270°, takes us to the point *D* below, and the fourth one will carry us round to *A*, where we started.

Then this circle is divided into four arcs, *AB*, *BC*, *CD*, and *DA*, each of which measures a right angle at the center.

RELATION OF THE EARTH TO THE HEAVENS 15

The right angle is subdivided into 90°, which may be done by dividing the arcs *AB*, etc., into 90 equal parts. Each degree is divided into 60 minutes, and each minute into 60 seconds.

Any little arc on the sphere is then said to *subtend* the angle formed between the lines drawn from the observer's eye to the ends of the arc.

To give an idea of the magnitude of angles, a foot rule at the distance of 57 feet from the eye subtends an angle of about 1°.

The diameters of the sun and moon subtend an angle of a little more than half a degree.

The diameter of the smallest round object that an ordinary eye can distinctly see subtends an angle of 1'. More exactly 1' is the angle subtended by a nickel at a distance of 320 feet.

7. The Relation of the Horizon to the Celestial Sphere. — In ordinary language the line around us where the earth and sky seem to meet is called the *visible horizon*, or simply the *horizon*. On a ship at sea the visible horizon is a circle, extending all round the observer. It is called the *sea horizon*.

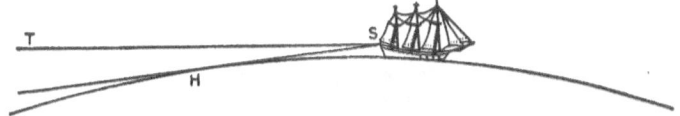

FIG. 4. — The dip of the horizon from the deck of a ship. The curve is the rounded ocean; *ST* is a horizontal line which does not strike the ocean at all; *H* is a point of the sea horizon, seen from the ship in the line *SH*. The angle *TSH* between the horizontal line from the observer's eye and the sea horizon is the dip of the horizon.

If we regard the earth as perfectly round and smooth, a plane resting on it at a point where we stand is called the *plane of the horizon*, or the *horizon plane*. As we look around, we must imagine this plane to extend out indefinitely on all sides of us.

In figure 4 we show the relation of the horizon plane to the sea horizon. The observer's eye being higher than the water, we see from this figure that the sea horizon will appear a little below the horizon plane. The angle by which it seems below is called the *dip of the horizon*. The higher the eye above the sea, the greater the dip.

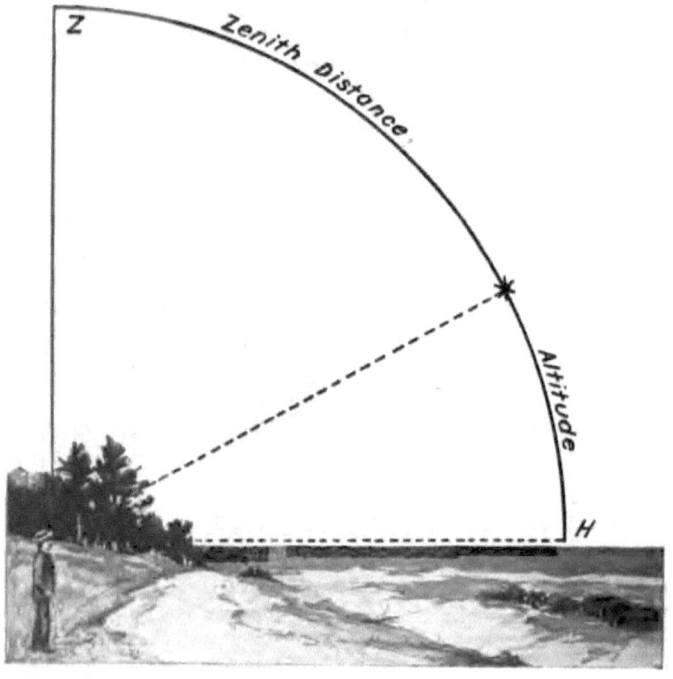

Fig. 5. — Showing the altitude of a body above the horizon. It is the angle between the body and the horizon *H* as it appears to an observer. The zenith distance is the angle between the line, in which the body is seen, and the zenith *Z*.

The point in the heavens over our head (*B*, fig. 3) is called the *zenith* ; the point in the celestial sphere below our feet (*D*, fig. 3), which we cannot see, is called the *nadir*.

The *altitude* of a heavenly body is the angle which its direction makes with the plane of the horizon. The greater its altitude, the higher it seems to us to be.

The line joining the zenith or nadir to the point where we stand is called a *vertical line*. It is evident that such a line is perpendicular to the horizon plane. Its direction is that of the plumbline.

The *zenith distance* of a body is its apparent angular distance from the zenith.

Zenith distance and altitude added together make up the whole arc from the zenith to the horizon, which measures 90°.

Change of Horizon as we Travel. — Now let us conceive the celestial sphere with the earth in its center. Let the globe in figure 6 represent the earth, and let APB be the horizon plane of an observer standing at P. We imagine this plane extending out all round until it meets the celestial sphere. We cannot draw this sphere round figure 6 because, to be large enough in proportion to the earth, we should have to make it bigger than a house. So we draw it on a smaller scale in figure 7. On this scale the earth is a mere point in the center.

The plane of the horizon at P in figure 6 cuts the celestial sphere in a circle AB, seen edgewise in figure 7. This circle is called the *celestial horizon* of the observer at P in figure 6. We must imagine it extending all around us, like the visible horizon. The celestial horizon divides the celestial sphere into two *hemispheres*, ACB and BDA.

Standing at P in figure 6, we can see the stars in the hemisphere ACB, figure 7, which is therefore called the *visible hemisphere*. We cannot see the stars in the hemisphere ADB because the earth is in the way, so this is called the *invisible hemisphere*.

We see in figures 6 and 7 that, as we travel from one place to another, the horizon changes its direction, so that stars, before invisible, will come into view on one side, and visible ones will seem to sink below the horizon on the other side. For example, if the observer travels to the point Q (fig. 6), his horizon plane will be *CD*, in figure 7. Then the stars in the region *N*, between *B* and *D*, will come into view, and those in the region *M*, between *A* and *C*, will seem to sink below the horizon.

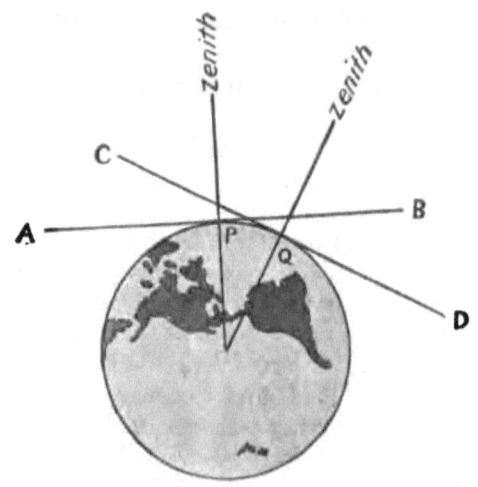

FIG. 6.—Showing how the horizon changes as an observer travels from one point of the earth to another. The straight lines *AB* and *CD* touching the earth represent planes seen edgewise. These are planes of the horizon, or horizon planes. Every different point of the earth's surface has a different horizon plane. For example: when the observer is at the point *P*, his horizon plane will be *AB*, which is represented as a line because we see it edgewise. If he travels to the point *Q*, his horizon plane will turn round into the position *CD*. Thus his horizon always changes as he travels. These horizon planes must be supposed extended out on all sides until they cut the celestial sphere. As the scale in this figure is too large to represent the celestial sphere, we make another figure on a much smaller scale to show the continuation of the horizon planes.

It is very interesting to watch this change during a long ocean voyage from the north to the south, or *vice versa*. As we steam along the rounded surface of the ocean, we see the constellations behind us sinking lower and lower every night, till they disappear below the ocean horizon, while new ones seem to rise from the ocean in front of us. Owing to the same cause the days are more than 24 hours long when we cross the

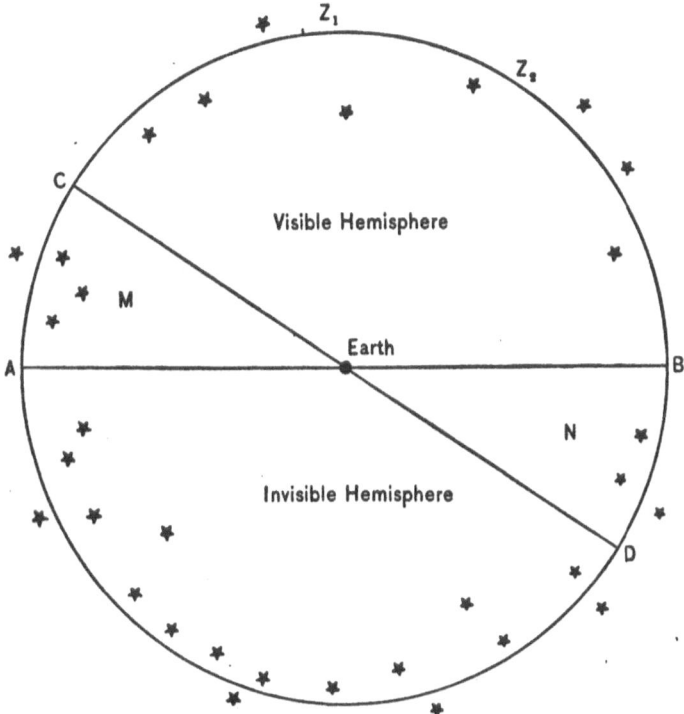

Fig. 7. — The celestial sphere, with the earth a mere point in the center at E, showing two horizontal planes corresponding to those of figure 6, extended until they intersect the celestial sphere. As an observer travels from P to Q, in figure 6, his horizon planes change from the position AB to the position CD, in figure 7. Thus the stars in the region M, which are visible when the observer is at P, are invisible when he is at Q, while the reverse is true in the region N. The position of the zenith on the celestial sphere also changes from Z_1 to Z_2, as the observer travels from P to Q.

ocean from Europe to America, and less than 24 hours long when we cross from America to Europe.

20 ASTRONOMY

8. The Diurnal Motion. — If, in our latitudes, we look at any star south of the zenith, and watch it for an hour, we shall find it moving from the left toward the right. If we watch it a few hours, we shall see that it moves yet farther and farther toward the right and at length sets in the west, like the sun. If we look at a star in the east, we shall find it rising upward and moving toward the south as the sun does every day. Thus the stars in the south rise and set like the sun.

We know that this is because the earth turns on its axis. To us, the appearance is the same whether we suppose the earth to turn and the stars to stand still, which is the truth, or whether we suppose the earth to be still and the stars to revolve round it, which is not the truth. In order to describe how things look, it is sometimes easier to suppose that the earth is still, as we may fancy it to be, and then to tell how the stars seem to move.

We call any seeming motion of the heavenly bodies, sun, moon, and stars, their *apparent motion*, which means their motion as it appears to us.

The apparent motion which they have in consequence of the earth turning on its axis, is called the *diurnal motion*. Hence, the motion by which the sun, moon, and stars rise and set every day is the diurnal motion.

9. Celestial Equator and Poles. — Figure 8 shows the earth with its axis passing through its center, from the south to the north pole. We must imagine this axis continued as far as we please in both directions.

Imagine a plane *EQ* passing through the center of the earth, at right angles to its axis. This is called the *plane of the equator*, because it intersects the earth's surface all round on the equator.

We must conceive this plane to be extended out as far as we please. Now we draw figure 9 on a small scale. The outer circle represents the imaginary celestial sphere with the earth as a black dot in the center.

The points *NP* and *SP*, in which the line of the earth's axis intersects the celestial sphere, are called the *celestial poles*.

The one which is visible to us, *NP*, is called the *north celestial pole;* the other, *SP*, which is below our horizon, is the *south celestial pole.*

The plane of the equator continued out in every direction intersects the celestial sphere in a circle *EQ*, which is called

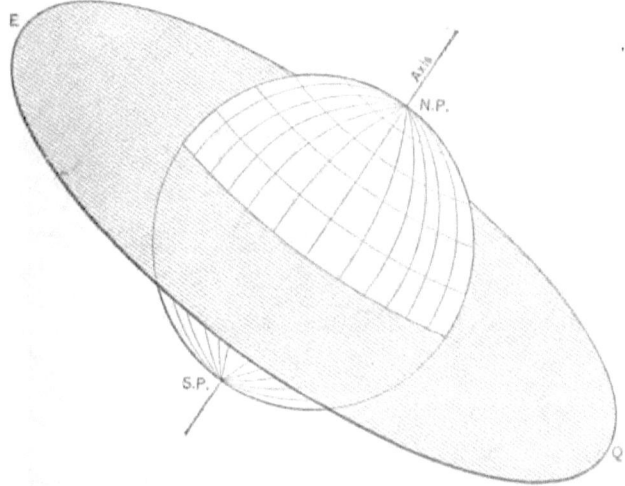

FIG. 8. — Showing the plane of the earth's equator. The equator itself passes around the earth and cuts the earth into two equal hemispheres. The plane of the equator extends outward on all sides until it meets the celestial sphere. In the figure we have to show it bounded by a circle because there is no room to represent it going any farther, but in reality it has no boundary.

the *celestial equator.* The celestial equator is an imaginary circle of the celestial sphere which always has the same position in the heavens. It intersects the horizon in the east and west points, and, in northern latitudes, passes south of the zenith by a distance equal to the latitude of the place. One half of it is above, the other half below, the horizon.

22 ASTRONOMY

10. The Meridian. — A line on the earth's surface, from the north to the south pole, is called a terrestrial *meridian* or simply a meridian. Owing to the curvature of the earth its form is that of a semicircle.

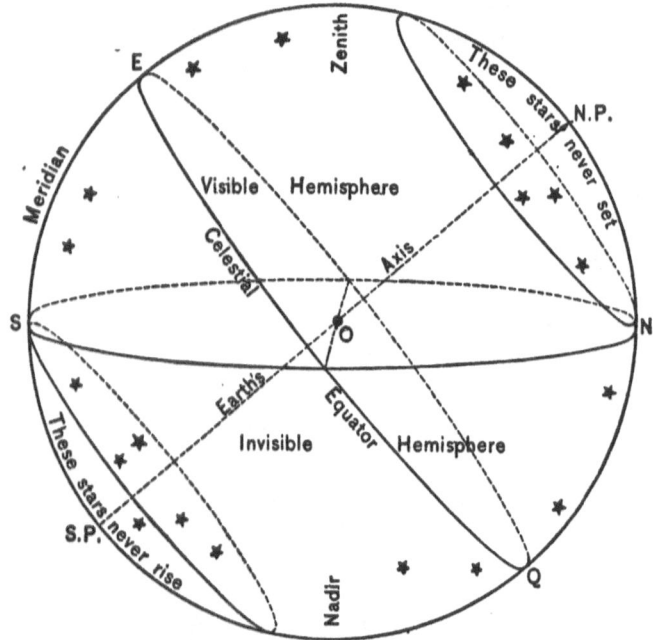

Fig. 9. — The celestial sphere, with the earth a point in the center at *O*, showing how some stars rise and set, while others never set, and others, in our latitudes, never rise. The circle of perpetual apparition, within which the stars never set, is round the north celestial pole, which is above our horizon. Those which rise and set are on both sides of the celestial equator.

The figure shows how the celestial meridian *SENP* passes over the celestial sphere in a north and south direction through the zenith.

Such a line may pass through any place: it is then called the meridian of that place. Thus we speak of the meridian of Greenwich, Washington, or Chicago, meaning the semicircles joining either of these points to the poles of the earth.

The north and south direction is that of the meridian. Any number of places may be on the same meridian; they are then north and south of each other.

The *plane of the meridian* of any place is a plane passing through the place and the north and south poles.

If we imagine this plane extended upward, so as to intersect the celestial sphere, the circle of intersection is called the *celestial meridian*. The celestial meridian passes through the celestial poles and the zenith of the place.

In figure 9 the plane of the paper is the plane of the meridian, and the outer circle of the figure is the celestial meridian. The points N and S in which it intersects the plane of the horizon are the north and south points of the horizon. Hence if one stands facing the south, his meridian rises perpendicularly from the south horizon to the zenith and continues to the celestial pole.

11. Diurnal Motion in Different Latitudes. — The apparent diurnal motion takes place as if the two celestial poles were pivots on which the celestial sphere is continually turning.

Suppose a person to stand at the north pole of the earth. Then the celestial sphere will appear to him as it is represented in figure 10. The north celestial pole will be over his head, that is, in his zenith. The plane of his horizon will be MN, which you see is parallel to the plane of the equator. When the earth becomes a mere point, these planes are so close together that we cannot distinguish between them. Hence: —

To an observer at the north pole, the north celestial pole is at the zenith, the celestial equator is in the horizon, and the south pole is at his nadir.

The poles being pivots, the diurnal motion of each heavenly body will seem to go on in a horizontal circle, from left to right, so that none of these bodies will either rise or set. When the observer is near the pole, the motion will be nearly horizontal.

Suppose the observer travels to latitude 40 degrees, which is nearly that of New York and Philadelphia. We have seen in figures 6 and 7 how his horizon plane turns round as he travels. As it turns, the celestial pole will appear to move over to a position where it will be 40 degrees above the horizon and 50 degrees from the zenith.

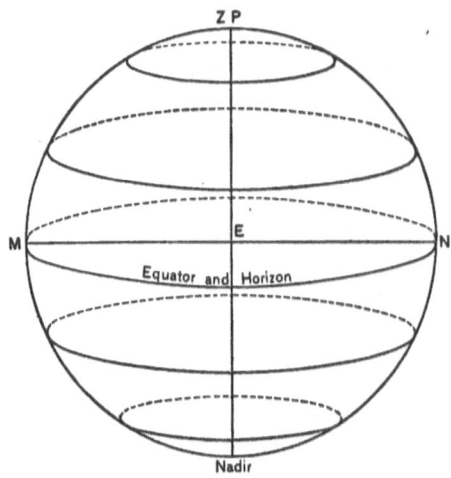

FIG. 10. — Showing the equator, horizon, and direction of the diurnal motion as seen by an observer at the north pole of the earth. The earth is an invisible point in the center at *E*. The outer circle is the celestial sphere as seen by the observer. Round his horizon on the celestial sphere is the celestial equator, which in this particular case is the same as the celestial horizon. The other horizontal circles show the diurnal motion of the heavenly bodies, which seem to make their revolutions round and round in horizontal circles, without either rising or setting.

In this position the northern heavens appear as in figure 11. The celestial pole is marked in the center of the map. There is a fairly bright star so near the pole that it is called the *Polestar*. If one has an exact north and south line, he can find the polestar by looking nearly halfway between the horizon and the zenith, toward the north. If not, he must find the constellation *Ursa Major,* commonly called the Dipper. The two stars which form the outside of the dish of the Dipper point very nearly at the polestar, as we see by the dotted line

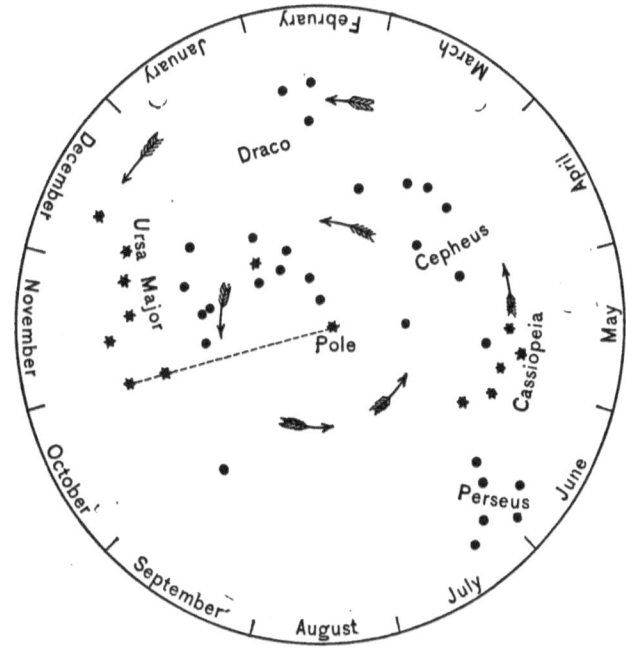

FIG. 11, — The circle of perpetual apparition to an observer in latitude 40°, showing the brightest stars of the northern constellations, including the pointers in Ursa Major pointing at the north star. To see how the stars will look at half-past eight o'clock on any evening in the year, hold the figure with the month at the bottom and look at the stars at the corresponding hour. The hour given is for the middle of the month. To find the positions at other hours, notice that the diurnal motion takes place in a direction the opposite of those of the hands of a clock, as shown by the arrows, and turn the figure accordingly.

in the figure. The Dipper can be seen almost any evening in the year, but in the evenings of autumn it will be low down near the northern horizon. If the pointers cannot be seen, there is only one other star that there is danger of mistaking for the polestar. This is *Beta Ursæ Minoris*, or Beta of the Little Bear, which is about 15 degrees from it and is of the same brightness; but it can still be distinguished by its being a little redder and by the two stars between which it is situated.

Having found the polestar, imagine a circle to be drawn in the heavens, having the pole as a center, and at such a distance

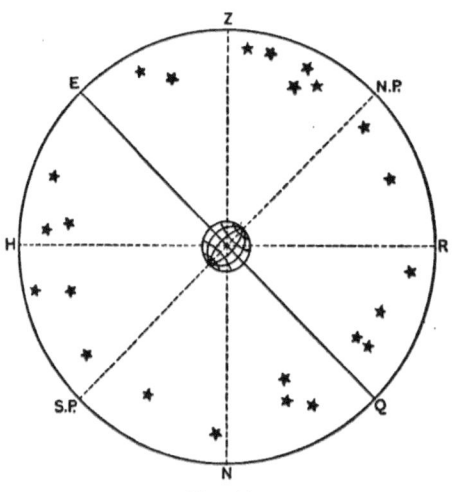

FIG. 12.

as to graze the northern horizon. Then studying figure 11, you will see that, as the celestial sphere appears to revolve around the pole as a pivot, all the stars within this circle will merely turn round and round the pole, as shown by the arrows, but none of them will ever rise and set. Hence this circle is called the *circle of perpetual apparition*.

It will readily be seen that the stars but a little outside this circle dip only a short distance below the horizon and are but a short time below it. They are above the horizon most of the

RELATION OF THE EARTH TO THE HEAVENS

time, but not all the time. The farther they are from this circle, the longer they are below the horizon.

Next let us study figure 12. This shows the celestial sphere as if we were looking at it from the east, so that we see the western portion of it. Here you see that the equator EQ is one half above the horizon HR, and one half below it. Hence

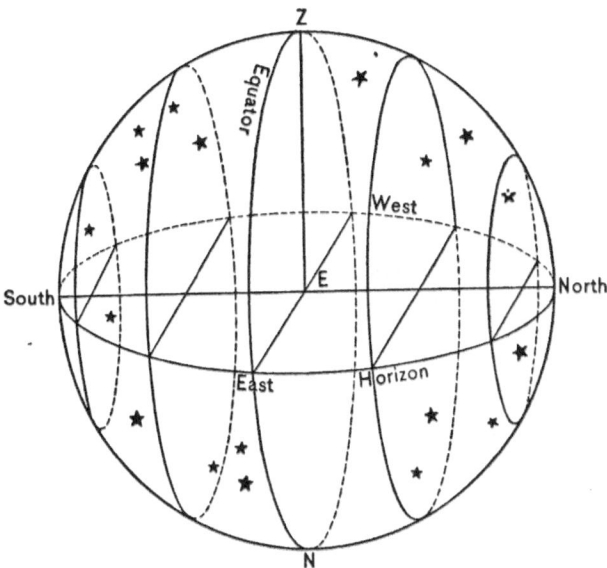

FIG. 13. — Showing the diurnal motion as seen by an observer at the equator. The earth is a point in the center at E. The celestial poles are in the north and south horizon, and the diurnal motion takes place in vertical circles shown in the figure. All the heavenly bodies are as long above the horizon as below it, except for a small effect of refraction, which will be hereafter explained.

a star on the equator is 12 hours above the horizon and 12 hours below it, in the course of each apparent diurnal revolution.

The farther south the star is situated, the shorter the time it is above the horizon and the longer the time it is below it. At

50 degrees south of the equator, the star will barely appear on the horizon and will immediately sink below it again.

Now notice the south celestial pole, as shown in figure 9. We see that, as the sphere seems to turn on it as a pivot, if we imagine a circle drawn so as to touch our southern horizon, the stars within this circle will never seem to us to rise at our latitude. Hence this circle is called the *circle of perpetual occultation*.

If we travel yet farther south, say to the Gulf of Mexico, the north celestial pole being nearer to the horizon, the circle of perpetual apparition will be smaller. The circle of perpetual disappearance will also be smaller.

If we travel to the equator, the two poles, or seeming pivots, being in the north and south horizon, all the stars will rise and set, each being as long above the horizon as below it. Hence there will be no circles of perpetual apparition or occultation.

If we go on into the southern hemisphere, the south celestial pole will rise above our horizon, the circle of perpetual apparition will be around it, and that of perpetual disappearance will be round the north pole, now below the horizon.

Because a meridian line is fixed on the earth's surface, it follows that all the meridians revolve with the earth. Hence one result of the diurnal motion is that all the heavenly bodies seem to pass the meridian every day. Really, the meridian passes them, but, to our senses, they seem to pass the meridian.

12. Right Ascension and Declination. — The *Declination* of a heavenly body is its apparent distance from the celestial equator. To measure it we imagine a circle to pass from the pole through the body S, figure 14, to the celestial equator. This is called an *hour circle*. The arc SR of this circle, between S and the equator, is the declination of the body S. It is measured in degrees, minutes, and seconds.

When the body is north of the equator, as at S, it is said to be in *North Declination;* when south, in *South Declination*.

RELATION OF THE EARTH TO THE HEAVENS 29

Comparing figures 14 and 9, we shall see that if we are in north latitude, the farther north the declination of a body, the longer it will be above our horizon during its diurnal revolution.

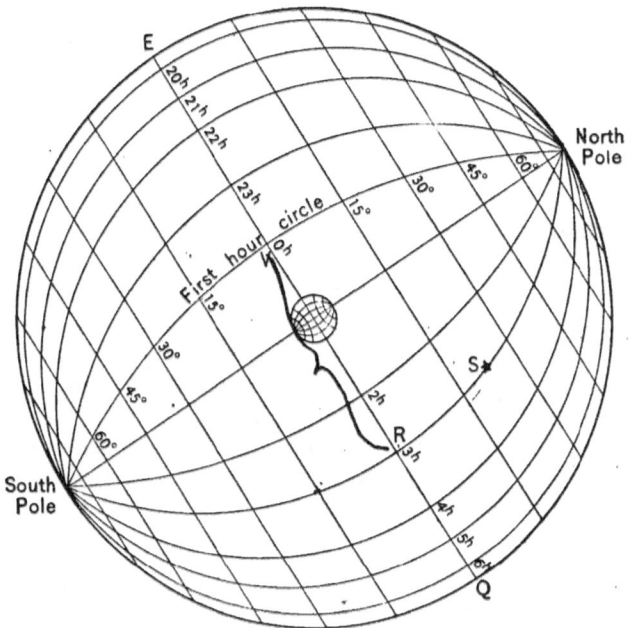

FIG. 14. — Showing the hour circles passing from one celestial pole to the other, and the right ascension and declination of a star. The first hour circle is represented as passing to the left of the earth through V. The declination of the star S is its distance from the equator R, and, in the figure, is about 23°. Its right ascension is the arc from V to R on the celestial sphere, which is here three hours.

The *right ascension* of a heavenly body is the angle which the hour circle drawn through it makes with that through the vernal equinox V; a point to be defined hereafter.

Astronomers commonly measure right ascension, like time, in hours, minutes, and seconds. As the earth turns through 360° in 24 hours, it turns 15° in every hour.

The position of a point on the celestial sphere — a star, for example — is expressed by its right ascension and declination, much as the position of a city on the earth's surface is expressed by its longitude and latitude. To understand right ascension we recall that the longitude of a place on the earth's surface is the angle between two terrestrial meridians, one of which passes through the Royal Observatory, Greenwich, and the other through the place. On maps are drawn as many meridians as are necessary, each being numbered according to its angle with the Greenwich meridian. The longitude of a place on the map is learned by its position relatively to the meridian lines which pass on each side of it.

On the same plan, astronomers suppose semicircles to pass on the celestial sphere from one pole to the other, as shown in figure 14. These circles are really celestial meridians. But the latter are supposed to move with the revolving earth, while the semicircles we speak of, that is, the hour circles, are fixed on the celestial sphere.

The first hour circle, taken as a standard, like the meridian of Greenwich, is that which passes through the vernal equinox. In figure 14 the vernal equinox is at V.

13. Correspondence of the terrestrial and celestial spheres. — On the earth a *parallel of latitude* is a circle parallel to the equator. All points on it have the same latitude.

In the heavens a *parallel of declination* is a circle parallel to the celestial equator. All points on it have the same declination.

The zenith of every point on the earth's surface lies on the corresponding parallel of declination. That is, if you are in 40° of north latitude, a star exactly over your head will be in 40° of declination. Thus, as shown in figure 14, every parallel of declination passes vertically over the corresponding parallel of latitude.

At any point on the earth's surface the altitude of the celestial pole is equal to the latitude of the place.

The arc of the meridian from the zenith to the celestial equator is also equal to the latitude of the place.

RELATION OF THE EARTH TO THE HEAVENS

To see this, suppose yourself standing at the equator. Then the celestial equator, as already explained, passes through your zenith, and the pole is in your horizon.

Now travel north through 1° of latitude. Then your horizon will have tipped 1° below the celestial pole, and your zenith will have moved 1° from the celestial equator. The corresponding result will occur how far soever you travel. In latitude 40° your horizon will have tipped 40° below the pole, so that the latter is now 40° above the horizon, and your zenith will have moved 40° from the equator. In latitude 45° the pole will be midway between the zenith and the north horizon; the equator will be midway between the zenith and the south horizon.

Algebraic Expression of the Relation between Declination, Zenith Distance, and Latitude. — Algebraic symbols are used to express these three quantities, the following notation being used: —

L = latitude, + when north; − when south.

D = declination; + when north; − when south.

Z = zenith distance; + when south; − when north.

If a star on the meridian is a certain distance Z south of the zenith, its declination will be less than that of the zenith by Z. At the zenith we have $Z = 0$ and $L = D$. Hence we shall always have $D = L - Z$, or $L = D + Z$.

Right ascension on the celestial sphere and longitude on the earth would correspond in the same way, were it not that the rotation of the earth keeps each meridian in constant motion over the hour circles of the celestial sphere. But every day there is a certain moment at which the vernal equinox passes over the meridian of Greenwich. At this moment the right ascension of a star on the meridian of any place is equal to the east longitude of the place from Greenwich. Hence, at this particular moment there is a correspondence of the meridians on the earth and in the heavens.

CHAPTER II

THE REVOLUTION OF THE EARTH ROUND THE SUN

1. The Earth as a Planet. — In the preceding chapter we have explained the various phenomena which arise from the rotation of the earth on its axis. We have to explain another change with which we are all familiar, — that of the seasons. We know that these go through a regular change in a period of about 365 days. During one part of this period, which we call summer, we see the sun rise to the north of east, pass the meridian high up in the heavens, and set to the north of west. At the opposite season the sun rises south of east, culminates low in the south, and sets south of west. In the first case the days are long and the nights short; in the second the days are short and the nights long.

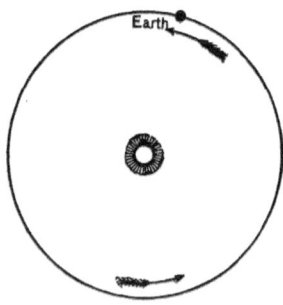

FIG. 15. — The earth moving round the sun.

Any one who thinks will see that this annual change of the seasons depends in some way on the sun; that the season is hot or cold, and the days long or short, according to the apparent path of the sun in the heavens. We have now to show that the changes of the seasons arise from the earth making an annual revolution around the sun. Thus the earth has two motions, its rotation on its own axis, and its revolution around the sun. The first produces day and night, the second summer and winter. To conceive the combined effect of these two revolutions is a task which requires some thinking. We have two things to consider, — the actual motion

REVOLUTION OF THE EARTH ROUND THE SUN

of the earth, and the apparent motion of the sun as we seem to see it.

If we could fly upward in a direction near that of the earth's axis to a distance of a thousand million miles, and then look back, we should see the earth and a number of other bodies forming, as it were, a little family far distant from all other heavenly bodies. The largest and brightest of those bodies would be the sun. At various distances and directions from the sun we should see eight or more smaller bodies looking like stars. If we watched long enough, we should see that those seeming stars were all in motion around the sun, each one keeping nearly, but not exactly, at the same distance from it during its course. The nearest would complete its circuit in about three months, while the most distant would take more than 160 years. These small starlike bodies are called *planets*.

The paths in which the planets perform their courses round the sun are called their *orbits*.

One of these planets is the earth on which we dwell. It is the third in the order of distance from the sun, and, as we have said, it requires a year to complete its circuit around the sun. It would be more exact to say that the time required for it to complete its circuit is what we call a year.

Besides the eight planets which we have described there are a number of smaller bodies going round the sun which we shall describe hereafter. This whole family of bodies is called the *solar system*. It is so called because the sun is the great central body on which all the others depend, and to which they all do homage, so to speak.

The distance of the earth from the sun has been determined in a number of different ways. According to the latest researches it is very nearly 93,000,000 miles; we scarcely know whether a little greater or a little less. Some idea of this distance may be gained by saying that a railway train running 60 miles an hour, and making no stop, would require more than 160 years to reach the sun. Five generations might be born upon it before the journey was completed.

The most marked difference between the sun and the planets is that the sun shines by its own light, while the planets shine only by the light that falls on them from the sun. Thus, so far as means of seeing are concerned, the sun is like a candle in an otherwise dark room and the planets are like little bodies seen by the light of the candle.

We have said that the bodies of the solar system form a group by themselves. Looking down from the height we have supposed, we should see this very clearly. The stars which stud the heavens would be seen just as we see them from the earth, in every direction. Their distances are so vast compared with the size of the solar system that even the latter, immense though it is, is but a speck in comparison. We may, if we please, call them suns. Most of them are brighter than the sun. They look small and dim because they are so much farther away.

Thus, having expanded our conceptions so that the earth shall be but as a point in the solar system, we must again expand them so as to think of the whole solar system as but a point in comparison with the distance of the stars.

2. Annual Motion of the Earth round the Sun. — We must now explain the motion of the earth in its orbit round the sun.

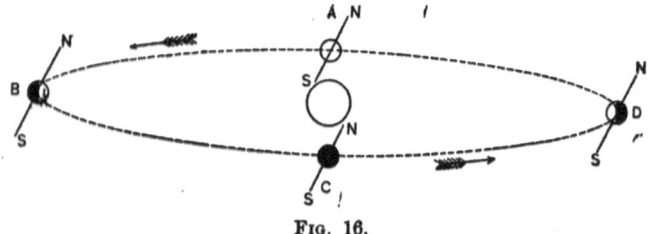

Fig. 16.

This is called its *annual motion*, because it takes a year to complete one revolution.

We cannot draw a figure which shall represent the earth and its orbit in anything like their true proportions, because

REVOLUTION OF THE EARTH ROUND THE SUN 35

the diameter of the orbit is more than 20,000 times that of the earth itself. So we have to draw figures, as before, on two very different scales. Figure 16 shows the orbit of the earth seen nearly edgewise. The plane containing this orbit is called the *plane of the ecliptic*. On a true scale the earth in this figure would be an invisible dot, so we make it larger, and then represent it on a still larger scale in figure 17.

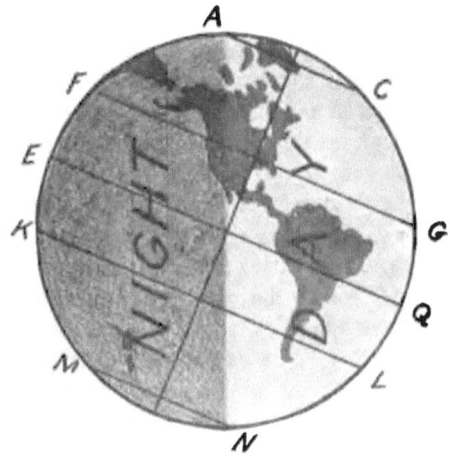

FIG. 17. — Showing how the sun shines on the earth in June, illuminating the whole of the Arctic circle, while the whole of the Antarctic circle is in darkness.

In figure 16 the direction of the axis is shown by the inclined line *NS*.

An important law of the earth's motion is this: **As the earth moves round the sun, the direction of its axis remains almost unchanged.**

This direction is not quite perpendicular to the ecliptic, but is inclined to the perpendicular by $23\frac{1}{2}°$, or a little more than one fourth of a right angle. This angle is called the *obliquity of the ecliptic*, because it is equal to the angle which the plane of the equator makes with the plane of the ecliptic.

3. How the Sun shines on the Earth at Different Seasons. — Let us now see how the sun shines on the earth at different times of the year.

Spring Position of the Earth. — About the 21st of March of each year the earth is in the position *A*, figure 16, where the line from the sun to the earth is at right angles to the earth's axis. The sun then illuminates the whole hemisphere of the earth which is turned toward it, from pole to pole. The days and nights are equal all over the earth. This time is called that of the *Vernal Equinox*, because the season in our hemisphere is spring, and the days and nights are equal.

Summer Position of the Earth. — Three months later, about June 21, the earth will be in the position *B*, with the north end of its axis now tipped toward the sun. The sun then shines on the region round the north pole, while that round the south pole is in darkness, as we see by figure 17, which represents the earth in the position *B*, but on a larger scale.

The circle *AC* round the north pole of the earth, which touches the edge of the illuminated hemisphere at this time, is called the *Arctic Circle*. Its radius will be $23\frac{1}{2}°$ of the earth's meridian, the same as the obliquity of the ecliptic. As the earth revolves on its axis in this position, the region within the arctic circle will never be carried outside of where the sun is shining. Hence, to an observer in this region the sun will not set on the 21st of June, but will seem to go round the sky in the direction from south through west, north, and east.

Next, imagine a circle, *MN*, drawn round the south pole of the earth, so as to touch the edge of the illuminated hemisphere. This is called the *Antarctic Circle*. We see that, as the earth revolves, the region within this circle will not be brought into sunshine at all. Hence the sun will never rise within this circle on June 21.

At a distance of $23\frac{1}{2}°$ north of the equator *EQ*, there is a circle *FG* on which the sun will be in the zenith at noon of June 21. This circle is called the *Tropic of Cancer*.

At the equator *EQ* the days will be equal to the nights.

REVOLUTION OF THE EARTH ROUND THE SUN

The further north we go from the equator, the larger the fraction of a circle of latitude round the earth which will be in sunshine. Hence on the 21st of June the days are longer and the nights shorter as we go toward the north.

South of the equator the days get shorter and the nights longer, as we travel south, until we reach the antarctic circle, when the sun will simply show himself on the horizon at noon.

Autumn Position of the Earth. — At C, the plane of the equator again passes through the sun, and the latter shines over one hemisphere of the earth, from the north to the south pole. At this time the days and nights are again equal the world over. This is called the *Autumnal Equinox*, because the days and nights are again equal and the season is autumn.

Winter Position of the Earth. — On December 21 the earth is in the position D, with the north end of the axis tipped away from the sun, and the south end tipped toward it. Now day and night are the reverse of what they were with the earth at B. Figure 17 will still answer for us, only it is now night where it is represented as day in the figure, and *vice versa*. All the region within the arctic circle is in darkness, all that within the antarctic circle in the sunshine. North of the equator the nights are longer than the days; south of it the days are longer than the nights.

The sun passes through the zenith of every place in latitude $23\frac{1}{2}°$ south at noon of this day. This circle of latitude is called the *Tropic of Capricorn*.

4. Apparent Motion of the Sun. — The Zodiac. Having explained how the earth turns on its axis and revolves round the sun, while, to us who live on it, it seems to remain at rest, we shall now explain how the sun seems to us to move. The apparent motion of the sun is based on these facts : —

1. Each fixed star is really in the same direction from us all day and all the year. The stars seem to us to change their direction only because we live on the moving earth.

38 ASTRONOMY

2. The sun is nearly, but not exactly, in the same direction from us all day, from its rising to its setting. But this direction changes during the year in consequence of the earth revolving round it.

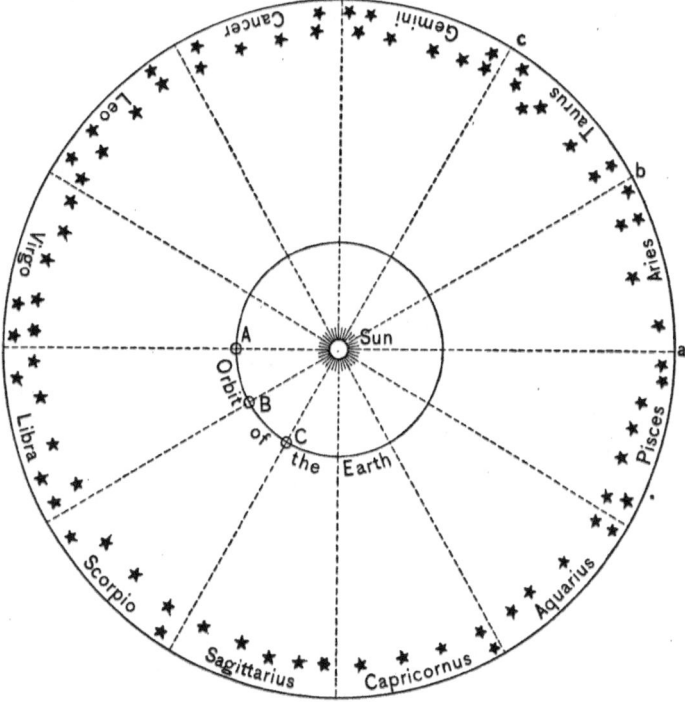

FIG. 18. — Showing how, in consequence of the earth moving around the sun, the sun seems to us to make an annual revolution round the celestial sphere among the stars, passing through the twelve signs of the zodiac.

Let us study figure 18, which shows the earth's orbit, *ABC*, with the sun in the center. Far outside the orbit lie the stars. To make a figure on the right scale, we should have to place the stars several miles away; as we cannot do this, we represent their positions as in the figure.

REVOLUTION OF THE EARTH ROUND THE SUN

Now suppose we could fly a few thousand miles above the earth and accompany it in its course round the sun. Then, looking down, we should see the earth turning on its axis, and bringing its oceans and continents into view, one after the other. Looking at the stars, we should see them at rest. They would neither rise nor set, nor even change their direction by any quantity we could perceive.

Next, let us see how it will be with the sun. When the earth is at the point A, we shall see the sun as if it were among the stars at the point a. A month later when the earth has got to the point B, the sun will appear among the stars at b. In another month, with the earth at C, the sun will be seen as if at c, and so on through the year. As the earth goes through its revolution round the sun, the sun appears to move around in a circle among the stars, until the earth gets back to the position A, when the sun will again appear in the position a. Hence : —

The sun appears to us to describe a complete circle around the celestial sphere, among the stars, every year.

The circle thus described by the sun on the celestial sphere is called the *ecliptic*.

The *zodiac* is an imaginary belt in the heavens, extending 8° on each side of the ecliptic, and passing all round the celestial sphere as the ecliptic does. The ecliptic is its central line.

If the axis of the earth were perpendicular to the ecliptic, the plane of the earth's equator would always pass through the sun, and the sun would always be seen in the celestial equator. Because of the obliquity of the ecliptic, already described, the ecliptic is inclined to the equator by an angle of $23\frac{1}{2}°$, cutting it at two points called the Vernal and Autumnal equinoxes, as shown in figure 19.

To make this clear, we show in figure 20 how, if we could see the stars around the sun, and the ecliptic and equator marked on the celestial sphere, we should, day by day, see the sun moving from west toward east, among the stars.

ASTRONOMY

In very ancient times men mapped out the apparent course of the sun round the celestial sphere, as shown in figure 19. They divided it into twelve parts, each 30° in length, and

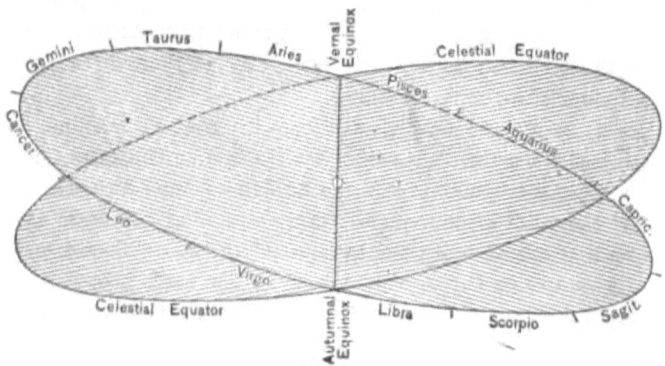

FIG. 19.— Showing how the celestial equator and the ecliptic span the celestial sphere among the stars, the two being inclined at an angle of 23½°.

Not only the earth, but the whole solar system, must be conceived as a point in the center of the figure. We must imagine ourselves looking out from this center. Then if we could see the stars around the sun we should see the latter appearing to pass around the ecliptic through the signs of the zodiac, as marked in the figure.

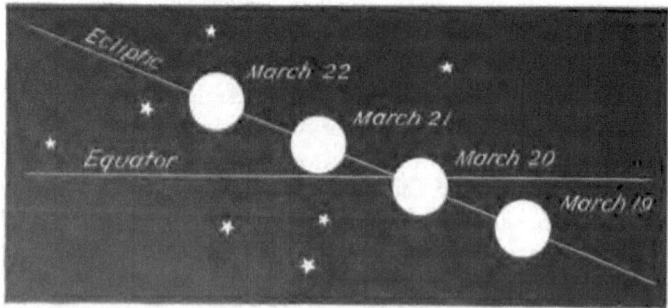

FIG. 20.— The sun crossing the equator about March 20.

named each part after the constellation in which the sun would have been seen had the stars been visible. These parts

REVOLUTION OF THE EARTH ROUND THE SUN

were called *signs of the zodiac*. The sun enters a sign about the 21st day of each month. The names of the signs and the months when the sun enters each are as follows: —

Aries, The Ram	March
Taurus, The Bull	April
Gemini, The Twins	May
Cancer, The Crab	June
Leo, The Lion	July
Virgo, The Virgin	August
Libra, The Balance	September
Scorpio, The Scorpion	October
Sagittarius, The Archer	November
Capricornus, The Goat	December
Aquarius, The Water Bearer	January
Pisces, The Fishes	February

When the sun is at the *Vernal Equinox*, it appears in the celestial equator, rises exactly east, and sets exactly west.

In figure 19 we see that during the six months the sun is passing from Aries to Virgo, it appears north of the celestial equator. It is therefore in north declination; it rises north of east and sets north of west. At this time, in the northern hemisphere, the days are longer than the nights. See the apparent diurnal course of the sun as shown in figure 22.

When the sun passes from Gemini into Cancer, it has reached its greatest north declination, and now begins to move south again. This point is called the *Summer Solstice*.

When the sun reaches Libra, it again crosses the equator toward the south. This point is called the *Autumnal Equinox*.

During the remaining six months, while the sun is passing from Libra to Pisces, it is in south declination; it rises south of east and sets south of west. In the northern hemisphere the nights are then longer than the days.

When the sun passes from Sagittarius into Capricornus it has reached its greatest south declination, and begins to return toward the equator. This point is called the *Winter Solstice*.

5. Seasons in the Two Hemispheres. — The reason that summer is hotter than winter is that the sun when north of the equator, not only shines longer upon us every day, but is nearer the zenith at noon. Thus more of its heat falls on any given surface — a square mile, for example, as shown in figure 21.

As the sun moves south in declination, its rays fall upon our portion of the earth at a greater obliquity, so that every square mile of our country receives less heat day by day.

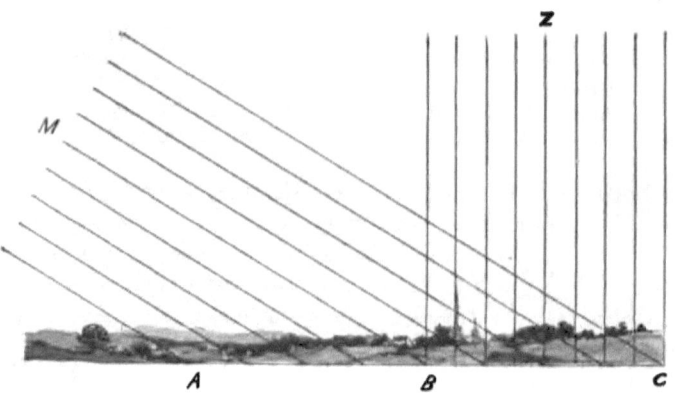

FIG. 21. — Showing how a square mile of the earth receives less heat, the nearer the sun is to the horizon. When the sun is in the zenith, the region *BC* receives as many of his rays as the region *AC*, twice as large, receives when the altitude of the sun is 30°.

The greatest amount of heat is received at the time of the summer solstice, about June 21, and the least at the winter solstice, December 22. But the highest average temperature does not occur till July. This is because the sun's rays require time in order to warm up the air and the surface of the land and sea, much as it takes time for a fire to warm up a room. The lowest temperature does not occur till January, because earth, air, and ocean retain for some time the heat radiated to them during the preceding months.

REVOLUTION OF THE EARTH ROUND THE SUN 43

But in the southern hemisphere the seasons are reversed. When the sun is in south declination, as at the winter solstice, the southern hemisphere has the longest days and the shortest nights. Hence, during our winter in the northern hemisphere, the southern hemisphere has its summer, and it has its winter during our summer.

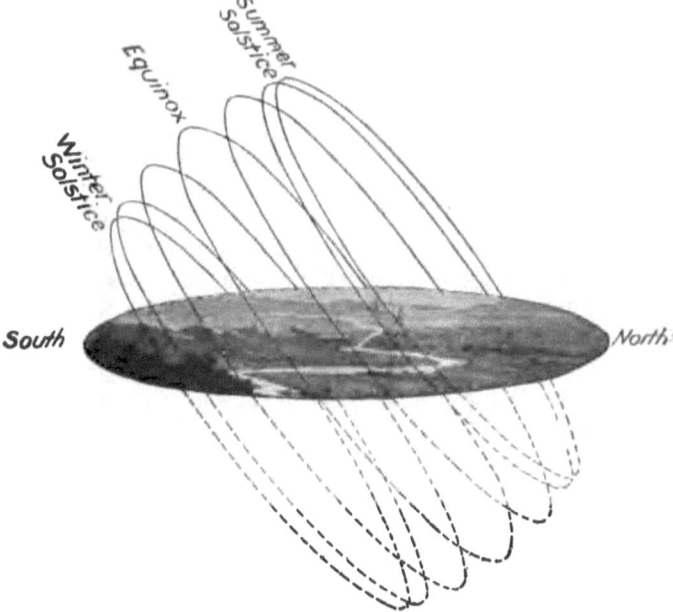

FIG. 22. — Showing the apparent diurnal course of the sun, as we see it in our latitudes at different times of the year. You must fancy yourself standing in the center of the landscape. Then in summer you will see the sun rise considerably north of east, pass not far south of the zenith at noon, and set to the north of west, as shown in the right hand circle of the figure. During the night it is completing that part of the circle which is below the horizon. During the remaining months of the year it seems to pass, day by day, farther and farther toward the south until December, when it seems to describe the left hand circle. It then rises south of east and sets south of west. We see that at this time the greater part of the circle is below the horizon, while in June the greater part is above the horizon.

We know that on a general average the hottest climates are within the tropics, and that the temperature is lower toward either pole. This is because the obliquity of the sun's rays increases toward the poles. At the poles the sun shines only half the year, and then is never more than $23\frac{1}{2}°$ above the horizon.

6. The Solar and Sidereal Years. — There are two ways of finding how long it takes the sun to complete its apparent revolution in the heavens, or, in other words, how long it takes the earth to make a complete revolution round it.

One of these consists in observing the exact time at which the sun reaches the equinoxes. In ancient times astronomical observers were able to do this by noting the days when the sun rose exactly in the east or set exactly in the west. By observing the rising and setting from day to day, they could find not only the day, but almost the hour in which the sun was on the celestial equator. Of course, with our more exact instruments, we can get this time with still greater precision.

The period between two returns of the sun to the same equinox is called the *solar year* or *equinoxial year*.

The other way of finding the length of the year consists in observing the interval of time between two passages of the sun past the same star in the heavens; for example, the period between two of its passages past one of the stars shown in figure 20.

This method seems to involve the great difficulty that we cannot see when the sun is near the star. But the astronomer has methods of knowing exactly where a star is by day as well as by night, and can determine the moment at which the sun passes it.

The ancient astronomers got the same result by using the moon as an intermediate object to measure from. The moon could be seen before sunset and its distance from the sun determined. Then, when the star appeared after sunset, the distance from the star to the moon was measured. Allowing

REVOLUTION OF THE EARTH ROUND THE SUN

for the motion of the moon during the interval, the apparent distance between the sun and the star could thus be learned from day to day. In this way it could be found how many days it was between the times at which the sun was at the same distance from any given bright star. This would be the period of apparent revolution of the sun in the celestial sphere, or, as we now know it to be, the period of one revolution of the earth in its orbit. This period is called the *sidereal year*, because it is fixed by the stars.

Hipparchus, who flourished about 150 B.C., was the first to make exact observations of the length of the year. Ptolemy, who flourished about 300 years later, made similar ones. They found that the length of the year, as determined in these two ways, was not the same, and that the solar year, as determined by the equinoxes, was several minutes shorter than the sidereal year determined by the return of the sun to the same star.

With our exact modern observations we have found the lengths of the years to be: —

Solar year,	365 d.	5 h.	48 m.	46 s.
Sidereal year,	365	6	9	9
Difference,			20 m.	23 s.

This difference shows that the position of the equinoxes among the stars is changing from year to year. Hipparchus and Ptolemy estimated the change to be about one degree in a century. We know it to be greater than this, — nearly one degree in 70 years.

7. Precession of the Equinoxes. — The motion of the equinoxes which causes the difference between the solar and sidereal year is going on all the time. It is called the *Precession of the Equinoxes*.

The nature of precession is now to be explained. The equinox is the point where the sun crosses the celestial equator. The position of the celestial equator on the celestial sphere is determined by the direction of the earth's axis, because the celestial equator is 90° from either celestial pole.

The precession of the equinoxes arises from the fact that the direction of the earth's axis in space is slowly changing.

Next, let us see how the change goes on. Imagine a line passing through the sun perpendicular to the plane of the ecliptic. The point in which this line, when continued to the stars, meets the celestial sphere, is called the *Pole of the Ecliptic*. It lies in the constellation Draco, the Dragon, but there is no bright star near it.

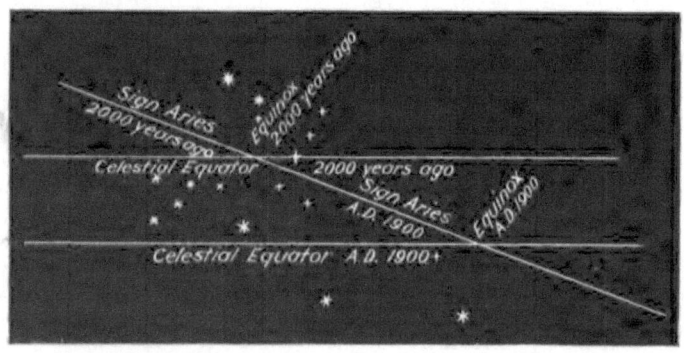

FIG. 23. — Showing how the equinoxes are gradually shifting in consequence of the motion of the celestial equator among the stars. One of the brightest stars in the figure, which was south of the equator two thousand years ago, is now north of it.

You will readily see that the angular distance between the pole of the ecliptic and the celestial pole, corresponding to the direction of the earth's axis, is equal to the obliquity of the ecliptic, $23\frac{1}{2}°$.

Now, the law of precession is that the celestial pole is in motion, and makes a complete revolution round the pole of the ecliptic in about 25,700 years. This motion is very slow to ordinary vision; it would take a century for the naked eye to notice it, even by careful observation. But the exact observations made by astronomers with the meridian circle make it evident month after month and year after year.

Owing to this motion of the celestial pole the celestial equator moves also, continually sliding along the ecliptic, and carrying the equinoxes with it, as shown in figure 23. This is why the equinox moves among the stars. The rate of motion is a little more than 50" in a year, or nearly 14° in 1000 years.

Motion of the Ecliptic. — If the plane of the ecliptic were absolutely fixed, the obliquity of the ecliptic would be always the same, and the motion of precession would go on forever at the same rate that it now does. But the attraction of the other planets on the earth produces a very slow change in the ecliptic itself, about $\frac{1}{50}$ the change of precession. In consequence of this change, the revolution of the celestial pole round the pole of the ecliptic does not take place at an exactly uniform rate, nor will it always be completed in exactly the same time. For the same reason the obliquity of the ecliptic slowly changes. It is at the present time diminishing at the rate of about 46" in a century.

Results of Precession. — One result of precession is that the celestial pole was not so near the polestar in former times as it is now. In ancient times it was so far away from that star that the latter could not be considered as a polestar at all. It has been continually coming nearer, and is still approaching it. About the year 2110 it will pass by the polestar at a distance of only 24'. Continuing its course, the celestial pole will pass some 5° from the star Alpha Lyræ, about 11,000 years from now, and will continue its circuit until it gets back to where it now is in about 25,700 years.

The two equinoxes will make a revolution round the equator in the same period of time, being carried along by the earth's equator, which is always at right angles to the earth's axis.

CHAPTER III

OF TIME

1. Diurnal Motion of the Sun and Stars. — We now know why it is that we do not see the same stars every evening all the year round. A star which, at any time, is seen in the west after sunset, will, evening after evening, be seen nearer and nearer the sun, until it is lost in the sun's rays. Then, when the sun has got considerably past it, we shall see it in the morning before sunrise.

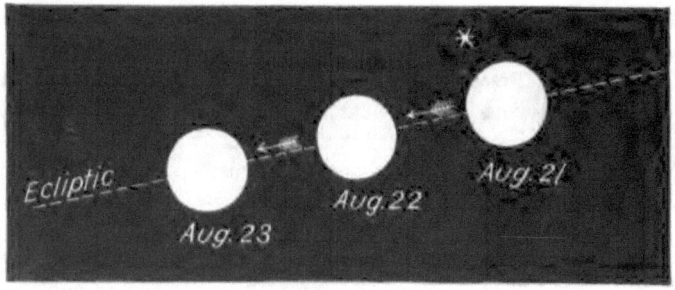

Fig. 24.

Imagine ourselves seeing the sun pass the meridian to-day. Suppose any star above it passing the meridian at the same moment. To-morrow, at noon, the sun will have moved a little east of the star (figure 24). Hence the star will pass the meridian before the sun does. Next day it will pass earlier than the sun by a yet greater amount, and so on through the entire year. At the end of the year they will again pass the meridian together. You see from this that the star, in its apparent diurnal revolution, has been continually running ahead of the sun and has caught up to it from behind at the end of the year. It follows that, in the course of the year, the star will have risen, crossed the meridian, and set one time more than the sun. The sun makes $365\frac{1}{4}$ apparent diurnal revolutions around the earth, there being one revolution a day. It follows that the star will have made $366\frac{1}{4}$ apparent diurnal revolutions.

If we divide the number of seconds in a day by $365\frac{1}{4}$, it will give us the time by which the star has gained on the sun every day. We find the quotient to be 237 seconds, or 3 m. 57 s. Subtracting this from 24 hours, we find the apparent diurnal revolution of the stars to be made in 23 h. 56 m. 3 s., or nearly 4 minutes less than a day. Hence, **any star passes the meridian three minutes and fifty-seven seconds, or nearly four minutes, earlier every day than it did the day before.**

The time between two successive passages of a star over the meridian is called a *sidereal day*, which means *star day*.

Astronomers divide the sidereal day into 24 sidereal hours, each hour into 60 sidereal minutes, and so on.

It will be seen from the way we have described the equinoxes that they each mark a certain point among the stars on the celestial sphere, and therefore that each equinox, like a star, crosses the meridian every day about 3 m. 57 s. earlier than it did the day before. When the vernal equinox crosses the meridian of a place, it is called *sidereal noon* at that place.

Sidereal time is the number of sidereal hours, minutes, and seconds since the vernal equinox crossed the meridian. It is counted from 0 h. to 24 h.

A *sidereal clock* is a clock so regulated as to keep sidereal time. Its pendulum is a little shorter than that of a common clock, so that it shall gain 3 m. 57 s. a day on the latter. The hands being set so that they shall read 0 h. 0 m. 0 s. as the vernal equinox crosses the meridian, the successive passages of the 24 principal hour circles over the meridian are told off by the clock. By looking at the clock, the astronomer can determine at any moment the position of any constellation relative to his meridian, and can point his telescope at any star he wants to see by day as well as by night.

2. Mean and Apparent Time; Inequality of Apparent Time. — The measure of time which we use in daily life is called *civil time*. The moment when the sun crosses our meridian we call *noon*. But there is a difficulty in using the true noon as 12 o'clock, owing to the obliquity of the ecliptic and the unequal motion of the earth round the sun.

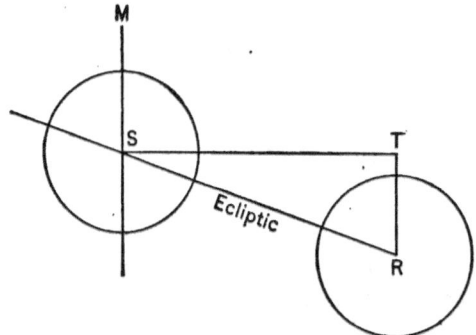

Fig. 25. — Showing the reason of the equation of time.

The earth moves a little faster in its orbit in our winter than it does in our summer. Hence the sun seems to move along in the ecliptic a little faster in winter than in summer. Owing to the obliquity of the ecliptic the earth sometimes has to turn farther in order that a meridian may catch up to the sun, than it does at other times.

Figure 25 shows this. Suppose that on some day at noon near the vernal equinox we see the sun at *R*. Next day the point *R* being fixed among the stars will pass the meridian 3 m. 57 s. before noon, and the sun will be at *S*, having moved obliquely toward the north. In order that the sun *S* may reach the meridian *TR*, it will have to pass over the distance *ST* by its apparent diurnal motion; or, in other words, the meridian *TR* will have to pass over the distance *TS* to be at the sun. But the line *ST* is shorter than *SR*. Hence it will take the sun less than 3 m. 57 s. to pass from *S* to *T*, so that it will be on the meridian a little earlier than it was the day before.

Next suppose the sun near the summer solstice. On account of the convergence of the meridians from the equator toward the north pole, the sun will pass over more than 3 m. 57 s. of right ascension near the solstices, and so will pass the meridian later each day than the day before.

In consequence of these two inequalities, the sun, at certain times of the year, falls behind, little by little, day after day, and at other times it catches up again, making the times between noons longer at some seasons than at others. Hence, if we used time measured by the true position of the sun, our hours would be of slightly unequal length. This unequal time, measured by the true sun, is called *apparent time*. The moment when the real sun is on the meridian is called *apparent noon*.

In former times, when people did not have good watches or clocks, and the exact time was not important to know, they generally went by the sun in setting their timepieces. But owing to the inequality of the intervals between two apparent noons, a timepiece will not keep apparent time.

Mean Time. — To make the hours of equal length, we fancy an imaginary sun to move round the celestial equator at a uniform rate, so that the true sun shall be sometimes ahead of and sometimes behind the imaginary sun. The latter is called the *mean sun*.

When the mean sun passes our meridian, it is called *mean noon*. Time measured from mean noon to mean noon is called *mean time*. This is the only kind of time we can measure with a clock, and it is the only kind now in general use.

Equation of Time. — The difference between apparent time and mean time is called the *equation of time*. It is greatest early in November of every year, when the true sun crosses the meridian about 16 minutes before mean noon. In February, the true sun is nearly as far ahead of the mean sun and crosses the meridian about 14 minutes after mean noon. Thus the greatest mistake we should make in measuring time by the true sun would be about a quarter of an hour.

Some almanacs give the equation of time for every day in the year, or, which amounts to the same thing, the time to which you should set your clock every day at the moment when the sun is on the meridian. The following examples will make this clear: —

    February 11, sun on meridian at 12 h. 14 m. mean time
    April 15, sun on meridian at 12   0   mean time
    May 14, sun on meridian at 11   56   mean time
    June 14, sun on meridian at 12   0   mean time
    July 26, sun on meridian at 12   6   mean time
    September 1, sun on meridian at 12   0   mean time
    November 3, sun on meridian at 11   44   mean time
    December 25, sun on meridian at 12   0   mean time

We see that there are four days in the year when the sun is on the meridian at mean noon, so that the mean and apparent time are then the same.

3. Local Time and Longitude. — As the earth revolves on its axis, all its meridians in succession pass the sun, or, as it appears to men, the sun passes all the meridians in its apparent diurnal motion round the earth. Because it is noon when the sun is on the meridian of a place, we see that noon is continually traveling round the earth, getting back to the same place in 24 hours. The circumference of the earth being 360°, we find, by division, that noon travels round the earth at the rate of

        15° in 1 hour of time
        15' in 1 minute of time
        15'' in 1 second of time

In the latitude of the middle states, 1' of longitude is about 4800 feet. Hence, 15'' is about 1200 feet. Thus we see that, in our latitude, noon travels from east to west at the rate of about 1200 feet a second. It requires between 4 and 5 seconds to travel a mile. Hence, two places do not have the same time at the same real moment unless they are north and south of each other.

Time measured at any place from the noon of that place is called *local time*. Hence the local time of two places a mile east and west of each other will differ in our latitude between 4 and 5 seconds. As there are 3600 seconds to an hour, it follows that at two places 800 miles east and west of each other, the difference of time will be about an hour.

4. Standard Time. — When people traveled by stage coaches or sailing vessels, and did not have very good watches, this difference of local time in different places caused them no trouble. When a man drove by stage to another place he set his watch on the new time. But when railroads got into operation, so that a man could, in the course of a day, travel from one place to another where the time was half an hour or more different, it was very troublesome. Nearly every railroad chose the time of one of the principal cities through which it passed, and thus it happened that travelers would frequently miss a train by mistaking the time.

To remedy this inconvenience, our system of standard time was introduced in 1883. Four standard meridians through the United States were chosen, 75°, 90°, 105°, and 120° west of Greenwich. By looking at a map of the United States we may see where these meridians run.

The eastern meridian, 75° west of Greenwich, passes through central New York, and a little east of Philadelphia, between that city and Trenton.

The central meridian, 90° west, passes through Wisconsin a little west of Madison, and down the Mississippi Valley through New Orleans.

The meridian of 105° passes along the eastern slope of the Rocky Mountains, through Wyoming, Colorado, and New Mexico.

The meridian of 120° passes through Washington, Oregon, and California, entering the Pacific Ocean on the coast of the latter state.

The local time of any one of these meridians is called *standard time*. To see how it compares with local time, as already defined, suppose a telegraphic operator anywhere on the 75th meridian to signal the exact moment at which mean noon reaches his meridian to all the people east of longitude $82\frac{1}{2}°$. When these people hear his signal they set their watches at 12 o'clock, so that they will all agree. The time their watches then keep is called *Eastern Time*.

One hour later, mean noon has reached the 90th meridian. At this moment we fancy a telegraph operator to send a signal to every one within $7\frac{1}{2}°$ of that meridian, that is every one living between $82\frac{1}{2}°$ and $97\frac{1}{2}°$ of longitude. These people all set their watches at 12 the moment they hear his signal. At this moment it will be 1 o'clock by all the eastern watches, so that the watches in question will be one hour behind the eastern ones. The time they keep is called *Central Time*.

The people in the Rocky Mountain region, including all those between $97\frac{1}{2}°$ and $112\frac{1}{2}°$ of longitude, wait another hour, till mean noon has reached the meridian of 105°, and then all, at the same moment, set their watches at 12. The time they keep is called *Mountain Time*.

In another hour mean noon has reached the meridian 120° west. At this moment all the people in the region west of $112\frac{1}{2}°$, which includes the whole Pacific Coast, set their watches at 12. The time they keep is called *Pacific Time*.

We see that

at 12 o'clock Pacific time
it is 1 o'clock Mountain time
and 2 o'clock Central time
and 3 o'clock Eastern time.

Hence the standard times differ only by one, two, or three hours. When one travels from San Francisco to New York, if

OF TIME

he wants his watch to keep the time of each region through which he passes, he sets it an hour forward as he enters each region. If he travels west, he must put it back as he enters each region. These jumps of one hour in standard time are arranged for our convenience, so that when we travel we shall not have to use a different time at every town.

Understand clearly the difference between *local* and *standard* time. The former changes constantly as we travel east or west, because our meridian is then constantly changing. Sunrise and sunset are given in local time, because they constantly travel from east to west around the earth. Hence, if a watch is set by the time of sunrise or sunset as we find it in an almanac, it will not give standard time unless we are on one of the standard meridians.

The difference between standard and local time is greatest at places half way between two standard meridians. Here the people can choose which meridian they will. If they choose the meridian next east of them, their watches will be half an hour fast of local time; if they choose that next west, they will be half an hour slow. Cincinnati, for example, is in longitude $84\frac{1}{2}°$. This is $5\frac{1}{2}°$ east of the central meridian, corresponding to 22 minutes of time. Hence, central time is there 22 minutes behind local time, so that noon at Cincinnati occurs 22 minutes before 12 o'clock central time

CHAPTER IV

OBSERVATION AND MEASUREMENT OF THE HEAVENS

1. Refraction of Light. — When rays of light enter a transparent substance, as water or glass, in an oblique direction, they are bent toward the direction of the perpendicular, as shown in figure 26. When they pass out of such a substance, they are bent away from the perpendicular, as shown at the bottom of the figure. This bending of light on entering or leaving a transparent medium is called *refraction*.

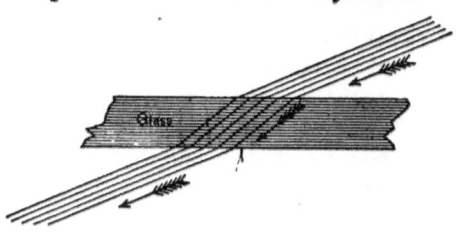

FIG. 26. — Showing the refraction of light in passing through glass, or any other transparent substance. When it enters the glass the rays are bent toward the perpendicular; when it leaves it they are bent from the perpendicular.

A familiar effect of refraction is seen in the bent appearance of an oar or a straight stick when held obliquely with its end under water. A clear pool looks shallower than it really is, for the same reason.

Atmospheric Refraction. — Refraction may be produced by a gas as well as by a solid body. If the gas is unequally dense in its various portions, a ray passing through it is refracted toward the denser portion. Now the air is more and more dense

OBSERVATION AND MEASUREMENT

Fig. 27. — Showing why a pool of water looks shallower than it really is in consequence of refraction. The looker-on imagines the bottom to be in a straight line in which he is looking, while in reality it is in the direction of a bent line and lower down.

from its upper limit to the surface of the earth. Hence, when a ray of light passes through it, it is bent by refraction, so that its course is concave toward the surface of the earth. This is shown in figure 28, which represents the earth surrounded by

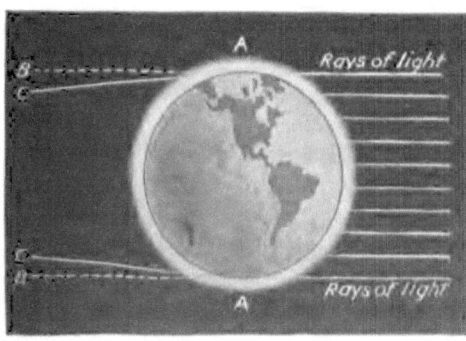

Fig. 28. — Showing how the sun's light is refracted by the atmosphere so as to illuminate the region within the earth's shadow. If the rays of the sun went in straight lines without refraction, they would pass through the point B, but, in consequence of refraction, they are bent down in the direction C. Hence, they meet each other before they reach the moon, and at the distance of the moon the whole interior of the shadow is illuminated with a lurid light.

its atmosphere. As a ray passes through the air, instead of continuing in the straight line AB, it is gradually bent so as to leave the atmosphere in the direction AC. This bending is called *atmospheric refraction*. When the light comes from a heavenly body, this refraction by the air is called *astronomical refraction*.

We always see a body in the direction from which the light it emits or reflects reaches our eyes. Thus, as a result of refraction, a heavenly body is always seen nearer the zenith than it really is. The refraction is small near the zenith, and at an altitude of 45° only amounts to 1′, a quantity that the eye could barely perceive. But it increases with great rapidity near the horizon, where it amounts to more than half a degree. Hence, on a level plain, or at sea, we still see the setting sun when its true direction is below the horizon, because at the horizon the refraction is a little greater than the diameter of the sun. In this case the lower limb of the sun is refracted more than the upper limb, which makes the sun look elliptical in form, the horizontal diameter being longer than the vertical diameter.

Another consequence is that the sun really illuminates a little more than half the earth, the curvature of the rays bringing them a little beyond where they would touch the earth if there were no atmosphere.

Dispersion. — If the surface where the ray leaves a homogeneous medium is parallel to that by which it enters, as in figure 26, the course of the ray after leaving will be parallel to its course before entering. But if the surfaces are not parallel, this will not be the case. If the transparent body is a triangular prism, as shown in figure 29, the ray may be refracted in the same direction both on entering and leaving. In this case ordinary light will not leave the medium as a single ray, but will be separated into rays of different colors. This separation of light into rays of different colors is called *dispersion*.

The effect of dispersion is seen very prettily when the rays of the sun are passed through a triangular prism of flint glass, and thrown upon a white screen or wall. We may then distinguish five very brilliant colors, red, yellow, green, blue, and violet, as well as some intermediate shades between these. If we notice how these colors are placed, we shall see that the

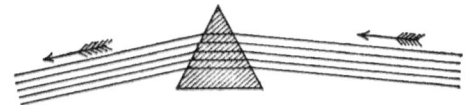

Fig. 29. — Showing how rays of light are refracted in passing through a prism.

red light is refracted from its course the least of all, yellow more, green yet more, and so on. This shows that the white light of the sun is a mixture of light of countless different kinds, each kind being refracted differently from the other kinds.

2. Lenses and Object Glasses. — When rays of light from a distant object pass through a convex lens, the curvature of the surface causes the rays to be more refracted the nearer they pass to the circumference of the lens. The result is that the rays coming from any one point of the object all converge very

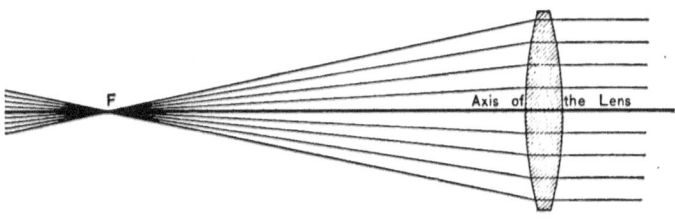

Fig. 30. — Showing how parallel rays of light are brought to a focus at F by passing through a convex lens. An observer holding his eye at F and looking at a light, however small, in the distance, would see the whole lens illuminated by the light.

nearly toward a certain point, F (figure 30), which is called the *focus* of the lens, and then diverge again as if they were emitted by the focus. The effect of this can easily be seen by holding a common reading glass or magnifying glass perpendicular to a window on the other side of a room. If you then hold a piece of white paper at the proper distance beyond the glass, you will see a little picture of the window on the paper. A picture thus formed by a lens is called an *image* of the object emitting the light that forms it. The lens may form the picture in the air when there is no surface on which the light may fall.

The *focal length* of a lens is the distance from its center to the image of a distant object formed by it.

The ordinary lenses which we use have convex surfaces and are called convex lenses. But a lens may be made having one or both surfaces concave. It is then called a concave lens.

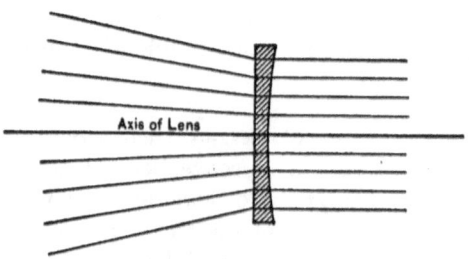

FIG. 31. — Showing how rays of light are made to diverge by a concave lens instead of being brought to a focus.

When rays from an object fall on a concave lens, they are not brought to a focus, but, on the contrary, are made to diverge by the refraction of the glass, as shown in figure 31.

An ordinary lens does not bring all the light actually to the same focus, on account of the dispersion of rays of different colors just described. The image of a star, instead of being a point, is a little colored circle near the focus. This dispersion is called *chromatic aberration*, and results in indistinctness of vision. But two lenses of different kinds of glass may be so formed that, when joined together, the rays passing through them shall all converge almost exactly to the same point. One of the lenses must be convex, the other concave. The convex lens is commonly made of crown glass, the concave one of flint. The property of these kinds of glass is that flint refracts light about as much as crown, but disperses the rays nearly twice as much. The dispersive powers of the concave and convex glasses act against each other, so that the rays leave the last lens without dispersion and so come to the same focus. Such a combination is called *achromatic* or free from color (figure 32).

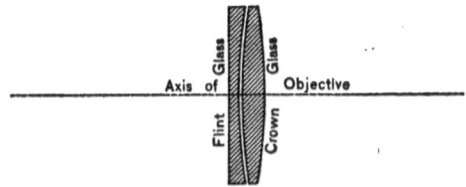

FIG. 32. — Section of an achromatic objective, showing the form of the flint and crown lenses. The crown lens is always convex; the flint has at least one surface concave.

3. The Refracting Telescope. — A refracting telescope is one in which the image is formed by a lens or achromatic combination of lenses called the *object glass* or *objective* of the telescope. When the telescope is pointed at a heavenly body or other distant object, the rays passing through the object glass come to a focus, and form an image of the object. This image is to be seen by the aid of an *eyepiece*, which is a combination of two small lenses so arranged that the observer can get as good a view as possible of the image.

Magnifying Power. — The magnifying power of a telescope is the number of times that it makes the linear dimensions of an object seem longer than they do to the naked eye. For example, the apparent diameter of Jupiter is commonly about 40''. A magnifying power of 50 would make it appear 2000'', or more than 33' in diameter, and larger than the sun or moon.

The law of magnifying power is that it is equal to the quotient of the focal length of the object glass divided by that of the eyepiece. Thus, with a telescope of eight feet focal length we should get a magnifying power of 96 by using an eyepiece of one inch focus, and a power of 192 by using one of half an inch focus.

It follows that, with any telescope, as high a power as we wish can be produced by using an eyepiece small enough. But a limit is soon reached beyond which a higher power will not enable us to see any more, because the light becomes fainter and the object more indistinct. Commonly an eyepiece between $\frac{1}{4}$ and $\frac{1}{2}$ an inch in focal length will show all that can be seen with any telescope.

As much of the sky as we can see in a telescope at one time (as magnified in the telescope) is called the *field of view* of the telescope. In the telescope, the field of view commonly looks very large, but the actual portion of the sky which it takes in is very small. The higher the magnifying power, the smaller it is.

The line which forms the central axis of the tube of the telescope, or which passes through the centers of the objective and eyepiece, is directed toward the center of the field of view. It is called the *line of sight* of the telescope.

4. The Equatorial Telescope. — If we point a telescope at a star, and do not move it, we shall see the star move rapidly across the field of view and disappear. This is because the telescope stands on the revolving earth, and turns with it. The apparent diurnal motion of a star, when seen in a telescope, is multiplied as many times as the telescope magnifies. Hence the higher the magnifying power of a telescope, the more rapidly the star will seem to move across the field of view. If we wish the telescope to stay pointed at a star, we must move it in the opposite direction to that in which the earth turns. This is done by supporting the telescope on axes on which it can revolve. The machinery by which the telescope is made to revolve, and the handling of the telescope made possible, is called the *mounting* of the telescope. A telescope mounted so as to follow a star in its diurnal motion is called an *equatorial telescope*, or simply an *equatorial*.

OBSERVATION AND MEASUREMENT

Figure 33 shows the mounting of an equatorial. In *P* is an axis parallel to the axis of the earth, the upper end of which, in the northern hemisphere, points to the north celestial pole. This axis is therefore oblique to the horizon. It is called the *polar axis* of the telescope.

To the upper end of the polar axis is fastened a sheath *D*, containing another axis. This is called the *declination axis*, because by turning the telescope on it, the latter may be pointed at any circle of declination.

By turning the telescope on these two axes we can point it to any part of the heavens. If we wish it to stay pointed at a star without our touching it, the telescope must be supplied with a clockwork

FIG. 33. — A small equatorial telescope.

so made as to keep the telescope turning from east toward west, exactly as fast as the earth turns from west toward east. Then by pointing the telescope at a star, and starting the clockwork, the star will remain in the field of view.

5. The Reflecting Telescope. — Rays of light from a heavenly body may be brought to a focus by a concave mirror as well as by a lens, as shown in figure 34. On passing through the focus

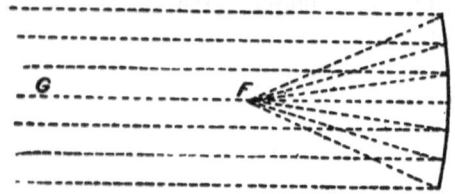

FIG. 34. — Showing how parallel rays falling on a concave mirror are brought to a focus at *F*.

they will diverge again, as they do after passing through the focus of a lens. Hence an image of a heavenly body may be formed in the focus of a concave mirror.

A *reflecting telescope* is one in which the image is formed by a concave mirror.

Such telescopes can be made of larger size than refracting telescopes, but they are not so convenient to use.

The observer, to view the image directly, would have to stand in front of the mirror, and thus be in the way of the light from the body to the mirror. The best way of avoiding this is to put a small diagonal reflector in the middle of the tube, near the focus, as shown in figure 35. Then the observer looks in sidewise near the end of the telescope where the eye is shown in the figure. The small mirror and its supporting piece cut off some of the light, but not so much as the observer's head and shoulders would cut off if he looked directly at the image.

6. Great Telescopes. — Large telescopes are objects of so much interest, that a short history of their growth will be given. The object glasses of the first telescopes, namely, those made by Galileo and his immediate successors between 1610 and 1750, consisted of only a single lens. Such a lens, as we have already seen, refracts the light of different colors to different foci. For this reason distinct vision was impossible with these

OBSERVATION AND MEASUREMENT

instruments. Vision was, however, improved by making them very long. It is said that some were 100 feet or more in length, but these proved to be quite unmanageable and were probably of very little use.

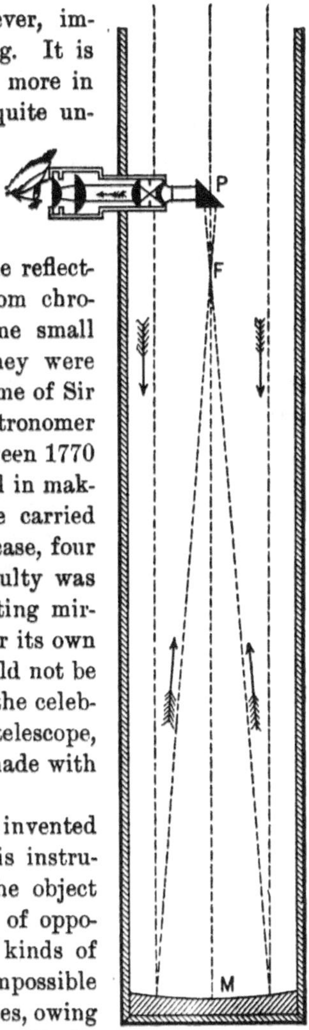

This difficulty with the refracting telescope led Sir Isaac Newton to propose the use of the reflecting telescope, which was free from chromatic aberration. He made some small instruments of this kind, but they were little more than toys until the time of Sir William Herschel. This great astronomer was at the height of activity between 1770 and 1800. He acquired such skill in making reflecting telescopes that he carried them up to two feet, and in one case, four feet in diameter. But the difficulty was then encountered that the reflecting mirror, when large, would bend under its own weight, so that a good image could not be formed. Thus, notwithstanding the celebrity of Herschel's great 40-foot telescope, his observations were nearly all made with smaller instruments.

About 1760, Dollond of London invented the achromatic telescope. In this instrument, as previously described, the object glass is composed of two lenses of opposite curvatures and of different kinds of glass. But it was long found impossible to make large achromatic telescopes, owing to the difficulty of making large blocks of flint glass of the necessary fineness and purity. This kind of glass contains a

FIG. 35. — A reflecting telescope on the Newtonian plan.

large quantity of lead, and the lead would sink down to the bottom of the pot in which the glass was melted, and thus make the glass heterogeneous and unfit for use. Thus, a hundred years ago, a refracting telescope four inches in diameter was considered large.

About 1810, Guinand, a Swiss glass maker, found a method of making disks of glass much larger than had before been possible. At the same time rose the celebrated Fraunhofer, a German optician, who acquired remarkable skill in grinding and figuring the lenses of object glasses into exact shape. He and his successors in Germany carried refracting telescopes up to 15 inches' aperture. In 1845 a telescope of this size was made for the Harvard Observatory in Cambridge, Massachusetts, and became, in consequence of its size and excellence, one of the celebrated instruments of the world.

About the same time, Lord Rosse of Ireland made his celebrated reflecting telescope, six feet in diameter. This is still, in size, the greatest telescope ever constructed. But the impossibility of keeping the mirror in proper polish and preventing it from bending under its own weight is such that more can be seen with smaller telescopes of different construction than with this celebrated instrument. Hence, it became desirable to improve the refracting telescope still farther.

The first one to improve on the work of Fraunhofer and his successors was an American, Alvan Clark, of Cambridgeport, Massachusetts. After making a number of small telescopes whose object glasses proved to be superior to any ever before figured, he succeeded, about 1861, in making a telescope of 18 inches' aperture, which was then the largest ever made. This instrument still exists at the observatory of the Northwestern University, Evanston, Illinois.

Shortly afterward, Cooke, of England, made a telescope of 25 inches' aperture, which was therefore much larger than that of Clark. This telescope was made for Mr. Newall of England. It is now at the University of Cambridge.

In 1873 Clark finished two telescopes an inch larger than

OBSERVATION AND MEASUREMENT 67

that of Mr. Newall. One was for the Naval Observatory of Washington, the other was given by its owner, Mr. L. McCormick, to the University of Virginia.

FIG. 36. — Great 40-inch telescope of the Yerkes Observatory.

Next, a telescope of 30 inches' aperture was made in France by the Brothers Henry, and is now mounted at the observatory of Nice on the coast of the Mediterranean.

In 1883 Mr. Clark and his two sons made the object glass of another telescope of 30 inches' aperture for the observatory at Pulkowa, in Russia. The mounting of this instrument was made by the Repsolds of Hamburg.

In 1876 Mr. James Lick, of California, gave money to found an observatory, which was to be provided with the largest telescope that had ever been constructed. The work of making the object glass of the instrument was again intrusted to Messrs. Alvan Clark and Sons, but great difficulty was found in getting disks of glass of the necessary size and purity. At length, after many years of failure, a Frenchman succeeded in the difficult task of making excellent disks of 36 inches' diameter. With these the Messrs. Clark completely finished the object glass of the telescope in the year 1887. The mounting was made by Warner and Swasey, of Cleveland, Ohio. Mr. Lick's observatory, which is called after him, was built on Mount Hamilton, in California, and the telescope commenced its work there in 1888.

The largest refracting telescope now in actual use is that built at the expense of Mr. Yerkes of Chicago for the university of that city. The object glass is 40 inches in diameter, and was figured by Alvan G. Clark, the son, and mounted by Warner and Swasey. The Yerkes Observatory, in which it is placed, is near the shore of Lake Geneva, Wisconsin.

7. Meridian Instruments. — A telescope of some sort is an essential part of every instrument intended for exact astronomical observation and measurement. One of the most common of astronomical instruments is the meridian *transit instrument*. Instead of being mounted like an equatorial, so as to be pointed in any direction, it turns on only a single horizontal axis, having an east and west direction. Thus the telescope turns only in the plane of the meridian, so that it will show us objects only while they are crossing the meridian.

To explain the use of the transit instrument we must recall what we have said about sidereal time. A sidereal clock is set

OBSERVATION AND MEASUREMENT

running in such a way that its hands shall point at 0h. 0m. 0s. when the vernal equinox is crossing the meridian. Then as the various heavenly bodies are seen in the transit instrument, crossing the meridian, the time shown by the hands on the face of the sidereal clock shows the right ascension of each.

If you should look into a transit instrument, you would see one or more dark lines passing up and down across the field of view. These are fine lines made of spider web, the middle one of which marks the meridian. The moment at which a star crosses this line may be noted on the clock within a small fraction of a second. This gives us the right ascension of the star with the same precision.

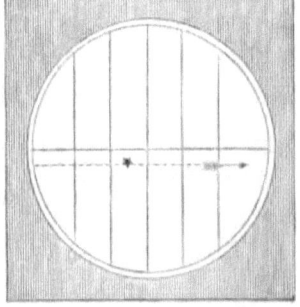

FIG. 37. — The threads in the focus of a transit instrument, with a star passing over them.

This instrument also enables us to determine the time of day, or the error of a clock or watch, with the same exactness. The observer notes the time by the clock at which a star of known right ascension crosses the meridian; the difference between the clock time and the right ascension is the error of his clock, for which he can make due allowance at any moment.

The moment of noon is sent out by a telegraphic signal from different observatories to railway offices and elsewhere, so that any one who receives the signal may set his clock exact to a second, if he has the skill to do it and exercises the necessary care.

To the transit instrument are sometimes attached vertical circles, which will turn with the instrument. These circles have fine lines engraved all round their circumference, so as to mark off the degrees and minutes of the circle. By their use the declination of a star, as it passes the meridian, may be observed.

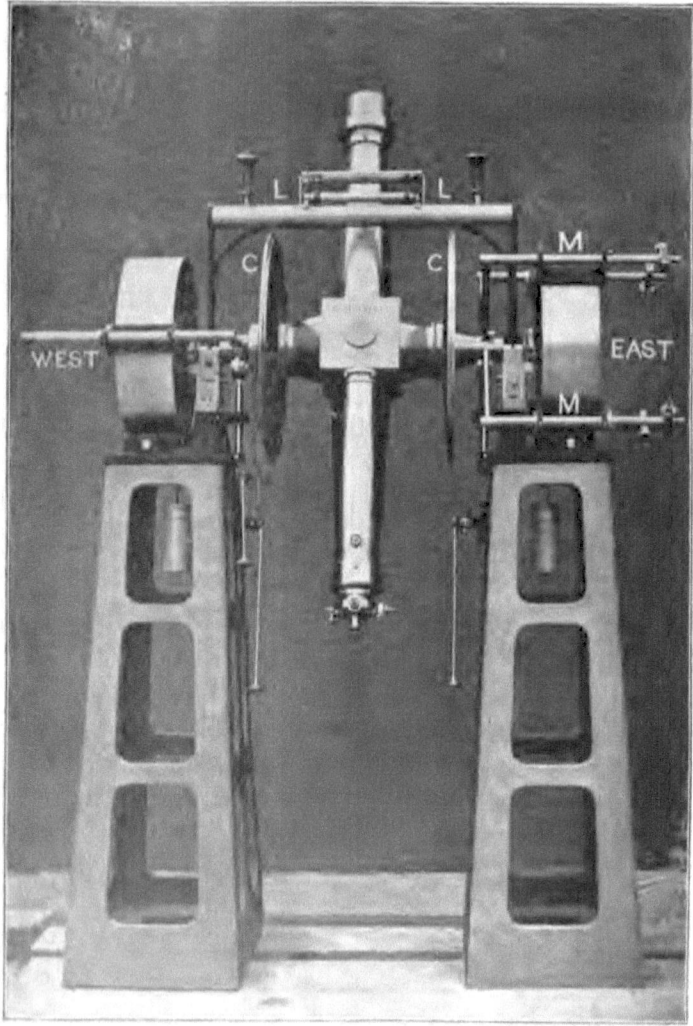

Fig. 38. — A meridian circle seen from the south. *C, C* are the graduated circles which are divided into degrees and small fractions of a degree, generally two or five minutes. *M, M* are four microscopes through which the graduations on these circles are read.

OBSERVATION AND MEASUREMENT

An instrument moving in the meridian, and provided with such a circle, so that both right ascension and declination may be observed, is called a *meridian circle*.

8. The Spectroscope and its Use. — We have seen that ordinary light, which seems to us white, is made up of countless rays of different colors, which can be seen separately by passing them through a prism, which causes them to be dispersed.

When the dispersed rays fall on a surface so that the different colors can be seen, the appearance is called a *spectrum*.

The *solar spectrum* is the spectrum which is formed when the rays of the sun are dispersed in the way described.

The *spectrum of a star*, a planet, or any other body, is the spectrum produced by the light coming from that body.

Different substances produce different kinds of spectra, so that it is frequently possible, from the nature of a spectrum, to know what kind of substance emitted the light, or through what kind of transparent medium the light has passed. The operation of detecting substances by their spectra is called *spectrum analysis*.

The Spectroscope. — A *spectroscope* is any instrument for showing or photographing a spectrum. The simplest spectroscope of all consists of a triangular prism of flint glass, like that of which the section is shown in figure 29. If we look through such a prism at a bright star or a distant light, we shall see the object spread out into a line of light of which the color varies from red at one end to blue or violet at the other, just as when the light of the sun passing through such a prism is thrown on a screen. When we look at the object directly, the retina of the eye is the screen on which the image falls.

The simplest kind of an astronomical spectroscope is shown in figure 39. This one is used in connection with a telescope. At *S* is a very narrow slit, between two plates of metal shown in figure 40. This is fastened to the eye end of a large telescope so that the slit shall be in the focus. This telescope,

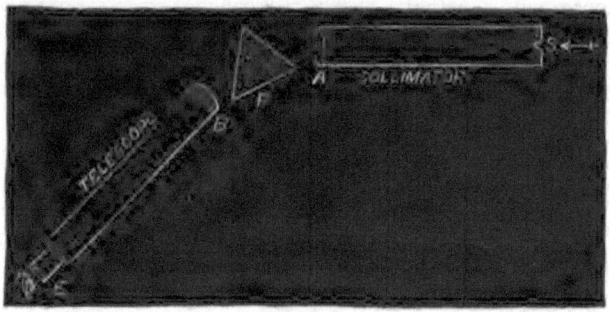

Fig. 39.

not shown in the figure, is pointed at a star so that the image of the star shall be formed in the slit through which all the rays from it will pass. Then the rays diverge as shown by the dotted lines and pass through an object glass A. The focus being at the slit, the rays of any color, yellow for example, will, after passing through the object glass, be refracted so as to be parallel to each other.

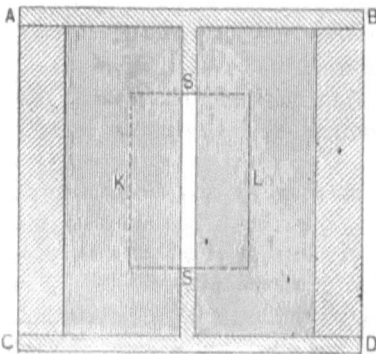

Fig. 40. — The slit of a spectroscope. AB and CD are slides in a plate of metal having an opening in it, shown by the dotted lines. This opening is nearly covered by the two plates of metal K and L, which move in the slides by a screw, and may be adjusted so as to form a slit SS as narrow as we please. This slit is placed in the focus of the telescope.

This combination of the object glass A with the slit in its focus is called the *collimator* of the spectroscope.

After leaving the collimator the rays next fall on the prism P, by which they are refracted in the way already shown. Next they pass to a second object glass *B*, which is part of a small telescope, and are again brought to a focus at *C*.

The action of the whole apparatus is such that rays of the same color come to the same focus at *C*, while those of a different color will come to a focus above or below the others, according to their different colors.

Then, an observer looking into *C* with an eyepiece *E*, as in an ordinary telescope, will see the spectrum of the star.

If, instead of using the eye, a sensitized photographic plate is inserted at *C*, a photograph of the spectrum may be taken. But this photograph will not, of course, show the different colors of the light.

In a powerful spectroscope, instead of a single prism at *P*, a long row of prisms is used in order to secure a greater dispersion of the light.

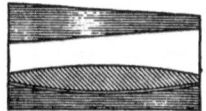

Fig. 41. — The objective spectroscope consisting of an object glass, with a refracting prism in front of it, by which the light of a star may be dispersed before it enters the telescope.

A yet simpler form of spectroscope consists of nothing but a telescope and a prism, the latter being put over the object glass as shown in figure 41. Then an observer looking into the telescope will see the spectrum of the star. A photograph of this spectrum may be taken in the same way as with the telescope. With such an instrument photographs of the spectra of all the stars in the field of view of the telescope may be made.

Kinds of Spectra. — If, with any form of spectroscope, we look at the flame of a candle or any other artificial light, not very far away, the spectrum will show the whole series of spectral colors without any dark lines; but if the light at which we look is several miles away, a distant gaslight, for example, we shall see that the spectrum is crossed by a number of dark lines. Why do these lines appear when the light is at a distance, and not when the light is near? Experiments show that it is because some of the light is absorbed by the air through which it passes. The light absorbed is that which belongs in the place of the dark lines. The process by which this is brought about is called *selective absorption*. The word *selective* implies that the air does not absorb all the different kinds of light equally, but picks out, or selects, certain kinds.

Spectrum of the Sun. — If, instead of looking at a distant light, we examine the light of the sun by the spectroscope, we shall find that, in addition to the lines produced by the air, there are a great number of other dark lines in the spectrum. These are formed by the selective absorption of the sun's atmosphere, which is different from the atmosphere of the earth.

If, instead of the sun, we examine the spectra of the stars, we shall find yet different lines. If we examine the spectra of some kinds of burning gas, we shall find one or more bright lines varying according to the nature of the gas. This shows that such a gas does not give out light of all colors, but only light of particular colors.

9. Semidiameter and Parallax. — If an observer with his eye at *E*, figure 43, is looking at a globe, as *AB*, the *apparent diameter* of this globe, as it appears to him, will be the angle be-

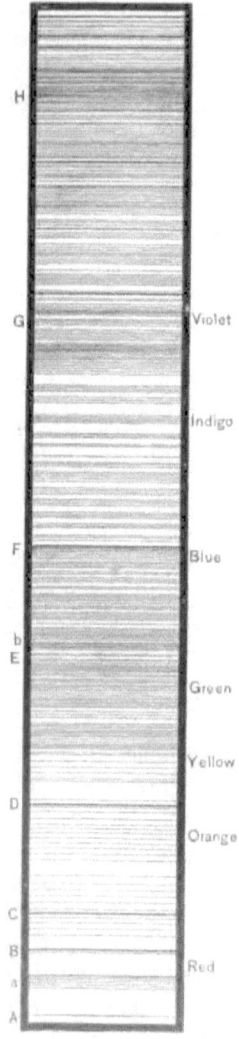

Fig. 42.
The solar spectrum with its dark lines.

tween the lines EA and EB drawn from his eye so as to touch the globe. One half of this diameter, or either of the angles AEC or BEC, is called the *semidiameter*. Thus by the

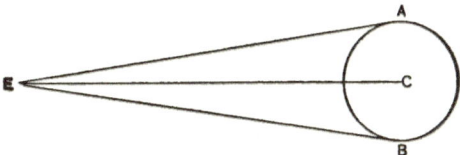

Fig. 43.—Apparent diameter of the sun, moon, or other heavenly body, as seen by an observer at E. The diameter is the angle AEB, subtended by the whole diameter of the body, while the semidiameter is the angle CBA or CBE between the center and apparent circumference.

semidiameter of the sun or moon we mean the angle between two lines, one of which is drawn to the center of the sun or moon, and the other to its apparent circumference.

It is evident that the semidiameter of a given body is smaller, the farther the body is away.

The *parallax* of a heavenly body is the difference of the directions in which it is seen from two different points.

Let S, figure 44, be the body, and A and B the two points from which it is seen.

Fig. 44.—Showing the parallax of a body at S.

An observer at A looking at S will see it in the direction ASX. An observer at B will see it in the direction BSY. The difference of these directions is the angle XSY, which is equal to the angle ASB. This angle is the parallax of the body for these two observers. In measuring parallax we may suppose the lines SX and SY continued till they reach the celestial sphere. The parallax is then an arc XY on the sphere.

Since the heavenly bodies are seen by observers from different points of the earth, they all have parallaxes. When an astronomer wishes to compute the direction of such a body from where he is stationed at a certain time, he first computes its direction from the center of the earth. Then he computes the difference of the direction from the center of the earth and from his point of observation. Since he is carried around by the turning of the earth on its axis, it follows that the parallax will be continually changing on account of this motion.

The *horizontal parallax* of a body is the difference of its direction AS (figure 45) when it lies in the horizon of an observer at A, and the direction CS in which it lies from the center of the earth. We see that this angle is the same as the semidiameter, ASC, of the earth as it would be if seen from the body S.

Fig. 45. —Horizontal parallax of a body at S. The circle represents the earth.

It is evident that the horizontal parallax of a body is less, the greater the distance of the body. Hence the heavenly bodies have a less or greater parallax according to their greater or less distances. Thus parallax and distance stand in a certain relation to each other.

The moon being the nearest body to the earth, it has the greatest parallax of all the heavenly bodies. Its horizontal parallax is generally about one degree. This is nearly twice its apparent diameter.

If an observer in New York, looking at the moon when near the meridian, should see her just south of a star, one looking at her in Chile would at the same moment see her north of the star.

OBSERVATION AND MEASUREMENT

The parallax of the other bodies of the solar system is so small that the eye could not perceive it. The sun's horizontal parallax is about 8.8''. This is the same thing as saying that the earth would have an apparent semidiameter 8.8'', if we could view it from the sun. Although a body of this size would be invisible to the naked eye, it would be a large object when measured with the telescope.

The only way in which the distances of the bodies of the solar system can be directly measured is by their parallax. Two observers on opposite sides of the earth, making exact observations of the direction in which they see the moon or a planet, can determine the parallax of the body observed, and from that can learn its distance. Thus, to express the distance of the sun, astronomers say that its horizontal parallax is 8.8''.

There are, however, other ways of getting at distances in the solar system. For example, the distance of the moon is determined by calculating how far off it must be to revolve around the earth in the time we see that it actually does revolve.

The distance of the sun is also determined by the velocity of light. It is found by experiment that light travels about 186,000 miles in a second. It is also found that it takes 500 seconds for light to travel from the sun to the earth. It is a very simple problem to find from these figures how far the sun must be.

10. The Aberration of Light. — It is found by very exact astronomical observations that we do not see a star in its true direction unless it lies in the direction of the earth's motion round the sun at the moment of observation. In all other positions the star will seem to be displaced in the direction toward which the earth carrying the observer is moving at the moment. For example, if the star is in the position S, figure 46, and the earth at E is moving in the direction of the arrow, then the star will appear as in the position T, in the direction shown by the dotted line. This displacement arises from the combination of the earth's motion with the motion of light. It is called the *aberration of light*, or, for shortness, *aberration* simply.

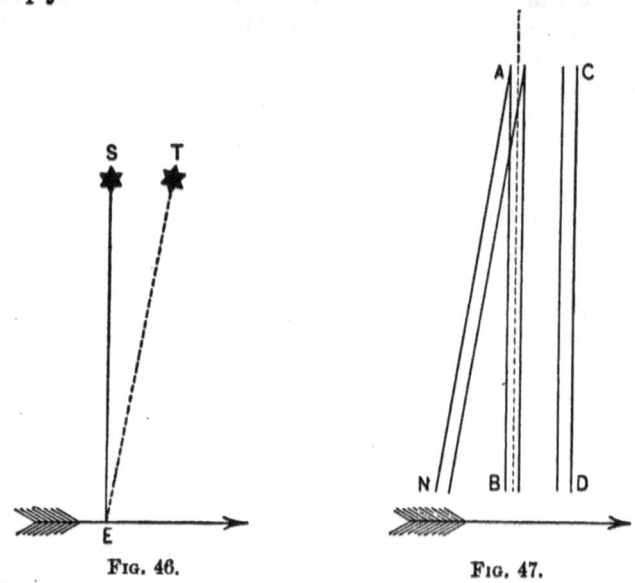

Fig. 46.   Fig. 47.

To explain aberration suppose *AB*, figure 47, to be a very long and narrow tube, and let the dotted line be a ray of light from a star, so that, if the tube were at rest, the ray would pass centrally through it, from *A* to *B*. Then an observer looking through the tube at *B* would see the star in the central line of the tube.

Now suppose the tube and observer to be moving in the direction shown by the arrow, so that while the light is passing from *A* to *B*, the tube is carried from the position *AB* to the position *CD*. Then, the motion of the tube would cause the ray to strike the side of the tube before getting through it, so that the observer would not see the star.

In order that the star may be seen while the tube is in motion, the latter must be inclined in the position *AN*. Then, while the ray of light is passing from *A* to *B*, the end *N* of the tube will be carried from *N* to *B*, and, in consequence, the light will pass centrally through the tube. Hence the observer, looking in at *N*, will now see the star in the direction *NA* instead of the true direction *BA*.

The velocity of light is very nearly 10,000 times that of the earth in its orbit. Hence the length *AB* is about 10,000 times the space *NB* through which the tube moves while the light is passing through it. The corresponding angle *NAB* is called the *constant of aberration*. Its amount is about 20.5″.

There are many every-day phenomena depending on the same principle that the aberration of light does. If a steamer while in motion has a side wind blowing against her, the wind will seem to the passengers to blow from a point nearer the direction in which the ship is going than the one from which it really blows.

If one drives rapidly through a shower of rain falling straight down, and watches the direction of the motion of the drops, they will be seen falling obliquely as if carried backward by a wind.

CHAPTER V

GRAVITATION

1. Force. — The motions of the heavenly bodies seem so different from the motions we are accustomed to see on the earth, that for many generations it was supposed that they could not be explained by the same laws. We have now to see how it is that the planets revolve around the sun according to the same laws which govern the motion of a ball thrown into the air. To do this, we must learn the meaning of certain words.

The substance of anything we can see or feel is called *matter*.

Anything made up of matter, and considered as a thing by itself, is called a *body*. For example, a ball is a body; the rubber, leather, and yarn of which it is made are matter.

That which makes a body move or stop moving is called *force*.

For example, if you throw a ball, your hand exerts a *force* on the ball; it is that force which sets the ball in motion. As the ball flies through the air, the air exerts a force against it; this force makes it go slower than it otherwise would go. When the ball strikes the ground, the ground exerts a force against it; this force soon stops it.

Friction is a force exerted by one body upon another that rubs against it. If you try to draw a sled on ice, you can pull it along very easily. But if you try to draw the same sled with the same load on a smooth pavement, you will have to pull harder, and on a rough pavement yet harder. This is

GRAVITATION

because the pavement exerts a greater friction against the runners of the sled than the ice does.

Gravity is the force by which all bodies on the earth tend to fall toward the center of the earth. This force is familiar to all of us from infancy. Every child that falls down and every stone that lies on the ground do so because of this force. If it did not exist, we could not keep ourselves or any loose object from slowly flying away from the earth except by fastening it down.

2. The Laws of Motion. — By long study and experiment men have learned that there are certain laws according to which all bodies move. These are called Newton's *laws of motion*. The first of these laws is this: —

Every body in motion, and not acted upon by any force, will move forward in a straight line with unchanging velocity forever.

It took men a long time to find out this law, because no unsupported body on the surface of the earth ever moves in a straight line, or continues moving forever. The reason is that all moving bodies around us are acted on by forces, in spite of all we can do to prevent their action.

One of these forces is that of gravity, which will always bring a body down to the earth if it is not supported. Another is friction. If a body is thrown through the air, the friction and resistance of the air are forces tending to stop it.

The less the friction, the farther a body will move. On a railway the friction is slight, so that a train will run some distance even if the locomotive leaves it. If there were no friction or resistance at all, it would run to the end of the road all by itself.

When a body is not held in any way, it is said to be *free to move*. No bodies on the surface of the earth are perfectly free to move unless they are thrown into the air, because, when they rest on the ground, there is always friction which hinders their motion. There is a little friction even if they float in water; and this friction will increase when the body is set in motion. But the heavenly bodies, not resting on or touching anything, are perfectly free to move.

The second law of motion is this: —

When a force acts on a body free to move, it takes time for the force to produce a change in the motion of the body, and the greater the force, and the greater time during which it acts on the body, the greater the change of motion of the body.

Every one must have noticed this when a train is starting from a station. When the engine first pulls, the motion is so slow that we hardly notice it. In a few seconds the train is going faster, and in a few seconds more, yet faster, the engine pulling its utmost all the time. A minute or more may be required before the strongest pull of the engine will get the train up to full speed. Then as people sometimes learn to their sorrow, it takes time for any force to stop the train. The engineer may apply his brakes with all their force, and yet the train move a quarter of a mile before stopping.

This property of matter by which time is required for a force to set it in motion, and again time is required for the force to stop it, is called *inertia*.

The third law of motion is this: —

Whenever one body exerts a force on another, the latter exerts an equal and opposite force on it.

Slap a wall with your hand, and you will find that the wall slaps your hand as sharply as your hand does the wall. When the wheels of an engine begin to turn, all the steam does is to make the wheels exert a force on the track. In doing this, the track exerts an equal force on the wheels, and this makes the engine move forward. When men floating on a raft in a shallow river push the bottom with their poles, the bottom pushes the pole, the pole the men, and the men the raft. This opposite force is called *reaction*, and the original force is called the *action*. Thus action and reaction are always equal, and in opposite directions.

3. Universal Gravitation. — Everything we see about us tends to fall downward toward the center of the earth. This is because the matter of the earth attracts everything on it toward the center. This attraction, or the tendency of things to fall downward, is called *gravitation*, and has always been known to men. What was not known till modern times is that all the heavenly bodies, sun, planets, and stars, also attract other bodies toward their centers. The general fact of this attraction is called *universal gravitation*.

When it was well understood that the planets revolved around the sun, it was still difficult for men to understand the laws according to which they moved. It was evident that there must be some cause connecting their motions with the sun. But when the laws of motion were not known, it was impossible to say exactly what that cause was. Between the years 1600 and 1680, it began to be suspected that there was some attraction between the sun and the planets. Sir Isaac Newton about the year 1680 was the first to prove this, and to lay down the laws of motion with such clearness and exactness that no doubt could remain on the subject. The law of universal gravitation, called Newton's law, is this: —

Every particle of matter in the universe attracts every other particle with a force which varies directly as the masses of the particles and inversely as the square of their distance from each other.

This is also called the law of the inverse square. It may be understood in this way: Twice as far away, the attraction is one fourth as much, 4 being the square of 2; three times as far away, one ninth as much, 9 being the square of 3; and so on. The distance of the moon is about 60 times the radius of earth; the square of 60 is 3600. Therefore, anything at the distance of the moon will be attracted toward the earth's center about $\frac{1}{3600}$ part as much as it would at the earth's surface.

It follows that the higher up we carry any body, the lighter it is. It is true that even in a balloon, the change of weight is not sensible to ordinary observation. But at the international office of weights and measures, near Paris, weighing has been brought to such perfection that, when one weight is laid on top of another in the scale pan, the combined weight of the two is found to be less than when they are laid side by side. The reason is that the upper weight is farther from the center of the earth when on top of the other than when lying alongside of it, and therefore is lighter.

The third law of motion applies to gravitation. Whenever one body attracts another, the other attracts it equally in return. A stone attracts the earth as much as the earth does the stone. A planet attracts the sun as much as the sun does the planet.

4. Weight and Mass. — The *weight* of a body is the force with which it is attracted by the earth.

The *mass* of a body is the quantity of matter which it contains.

The mass of a body is measured by its inertia, or by the force required to give it a certain motion in a certain time.

In ordinary life, mass and weight, at any place, are always in the same proportion to each other, so that we need not make any distinction between them. But in astronomy, where we have to consider bodies in the heavens, the case is very different.

Suppose we found the weight of a ham here on the earth, as determined by a spring balance, to be 24 pounds. Suppose we could then fly up to the moon with the ham. The moon has so much less matter than the earth that it attracts a body at its surface with only about one sixth the force that the earth attracts the body at its surface. Therefore the ham on the surface of the moon would weigh only 4 pounds in the balance. But there would be just as much ham there as there was on the earth, as one would find out on trying to eat it. Therefore the mass of the ham would be the same as before, although the weight was so much less.

GRAVITATION

If instead of using a spring balance we used ordinary weights, we should find the same weight on the moon as on the earth, because the weights would be lighter in the same proportion as the ham is. Therefore, to get the real weight, we suppose a spring balance to be used.

If a baseball club could fly to the moon and there play a game, the distinction between mass and weight would be very evident. The mass being the same as here, the pitcher would not be able to throw the ball any faster than he can throw it on the earth. The catcher would find the ball striking as heavy a blow in his hands as it does here, and the batsman could bat it no faster than he does here.

As the ball would be drawn toward the moon by only one sixth the force that the earth draws it, it would be found to be as light as a rubber ball. It would stay long in the air when batted, and home runs would be made all the time.

As the quantity of matter in a body is always the same, while the weight varies according to the attraction of other bodies, astronomers do not speak of the *weight* of heavenly bodies, but only of their *mass*. All bodies having the same mass attract equally at the same distance. Hence the mass of a body may be determined by the attraction between it and another body at some fixed place. If we could cut a planet into pieces small enough to be brought to the earth and weighed, we could determine the mass of the planet by the weight of the pieces. As we cannot do this, we determine the mass by the attraction which it exerts on a satellite or on some other planet.

5. How the Attraction of the Sun keeps the Planets in their Orbits. — In consequence of universal gravitation all the heavenly bodies attract each other. The sun is so very large and massive that it attracts the planets much more strongly than they attract each other. We have now to see that the planets move round the sun according to the same laws that govern the motion of a ball thrown by the hand.

Suppose the ball thrown in the line *AB*, figure 48. If there were no gravitation, it would, in a certain time, say one second, reach the point *B*, and in two seconds the point *D*. In consequence of gravitation it describes a curve *ACE*, as every one knows who watches the motion of a ball. The distance *BC*, or the drop of the ball from its line of throw in one second, will be 16 feet, no matter whether thrown slowly or rapidly. This is the distance which a body dropped from the hand will fall in one second. In two seconds the drop of the ball will be from the point *D* to *E*. This is four times as far as the drop in one

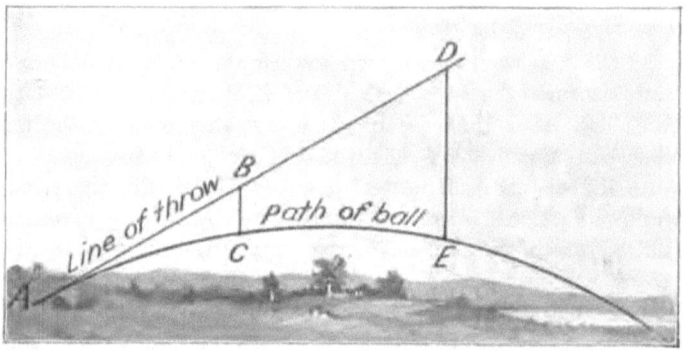

FIG. 48. — Showing the curved path of a ball thrown in the air and falling to the earth under the attraction of gravitation.

second, because the attraction of the earth is constantly acting on the ball, and increasing its velocity, thus making it fall farther every second than it did the second before. The law is a drop of 3×16 feet during the second second, of 5×16 feet during the third, etc. Adding up the falls in each second, we see that the total drop will be 4×16 feet in 2 seconds, 9×16 feet in 3 seconds, and so on, as the square of the time.

A little study will make it plain that the faster the ball is thrown, the less the bending or curvature of its path, because it must go farther before it will get the same drop. Thus the course of a bullet seems almost straight, while a ball thrown

GRAVITATION

by a little child describes a path much more curved than one batted by a baseball player.

Imagine the earth to be a body thrown like a ball, and attracted by the sun. To learn how it will move, we must note that although the mass of the sun is 333,000 times that of the earth, yet it is so far away that its attraction is only about $\frac{1}{1650}$ that of the earth on things about us. Hence a ton weight,

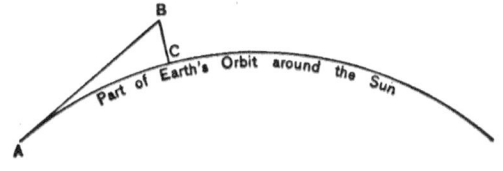

FIG. 49. — Showing a small part of the earth's orbit round the sun. In consequence of the sun's attraction, it is continually falling away from the line of its motion, AB for example. Compare this figure with the preceding one, and note that as a ball's path continually curves toward the earth, so the earth's path continually curves toward the sun.

or 2240 pounds, is here on the earth attracted by the sun with a force of little more than one pound.

To find how far a body like the earth would fall toward the sun in one second, we must divide the distance, 16 feet, or 192 inches, by 1650. This is less than $\frac{1}{8}$ of an inch. Now in figure 49 let the arc be a piece of a circle around the sun. Draw the line AB, touching the circle, and let the earth be thrown in the direction of this line with such speed that the curvature of the path in consequence of the fall of the earth toward the sun shall be equal to the curvature of the circle. Then, notwithstanding that the earth has commenced to fall toward the sun, when it reaches C, it has kept in this circle and is no nearer to the sun than it was at A, but is now going in a slightly different direction. In the same way, when it has described another arc, it has got no nearer the sun by its falling, but has only kept in the circle. All that the sun has done by its attraction is to keep the earth from flying off from it altogether in a straight line. It keeps bending the path of the earth from a straight line into the circle round the sun. Thus, instead of either falling into the sun or going away altogether, the earth revolves round and round the sun forever.

The idea we have now to grasp is that the earth is not held by anything, but is flying through space, turning on its axis all the while, with us upon it, as a ball might fly through the air with insects on it.

One who knows enough of geometry to be able to make the necessary calculations will find that in order that the earth may thus describe a circle round the sun, the speed with which it is to be thrown must be about 18.6 miles per second. That is, at the distance of 18.6 miles along the line AB the circle round the sun will be $\frac{1}{8}$ of an inch from this line.

But the earth does not always go exactly with this speed. We must, therefore, show what happens if the velocity should be a little less than that we have supposed. In such a case, the earth or other planet will fall a little nearer the sun until it gets halfway round. But, in thus falling nearer the sun, it will have acquired a greater velocity, and, in consequence of this increase of velocity, it will, after going halfway round, begin to recede from the sun until it gets back to the place it started from. Thus it will go round and round the sun, describing an orbit a little nearer the sun on one side than it is on the other. That is, the sun is not exactly in the center of the orbit. In another chapter we shall see that the orbit is an ellipse.

GRAVITATION

6. Centrifugal Force. — Let figure 50 represent a rapidly revolving wheel. To show the matter clearly, we suppose the rim to be cut into eight pieces by the black lines, so that each piece is fastened only by the spoke. Then at every instant, in virtue of the first law of motion, the parts of the rim will tend to fly off in straight lines in the direction in which they are at the moment moving. This direction is shown by the arrow heads.

But they are kept from thus flying off by the pull of the spokes upon them. By virtue of the third law of motion each part pulls on the spoke with the same force that the spoke pulls on it.

This pull is called *centrifugal force* because its direction is away from the center on every side.

The swifter the motion, the greater the centrifugal force. If

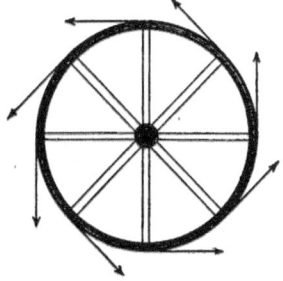

FIG. 50. — Centrifugal force.

the speed is increased without limit, the force will become so great as to break the spokes. Then, each piece of the rim will fly away in the straight line in which it is at the moment moving. A fly wheel regulating the motion of heavy machinery is sometimes known to break from this cause, and the pieces flying away through the roof of the building and the air may cause great damage.

The rotation of the earth on its axis causes a slight centrifugal force, which is overcome by gravity. One of its effects is to make all bodies on the earth's surface a little less heavy than they would be if the earth did not rotate.

Another effect is to make the earth and planets assume the form of oblate spheroids, as we shall explain, for the earth, in the next chapter.

CHAPTER VI

THE EARTH

1. Figure and Magnitude of the Earth. — If the earth did not rotate, the attraction of every particle of the matter composing it upon every other particle would tend to bring it into the form of a sphere. But the rotation of the earth on its axis generates a centrifugal force which partially counteracts the attraction of gravity at the equator, and thus makes the earth bulge out at the equator, so as to take the form of an oblate spheroid. In this form of spheroid the equator is a circle, and the axis or diameter through the pole, called the polar axis, is shorter than that through the equator.

The ratio in which the polar axis is less than the diameter at the equator is called the *ellipticity* of the earth. Its amount is about $\frac{1}{300}$; perhaps a little greater. That is, if we represent the equatorial diameter by the number 300, the polar diameter will be about 299.

The elevation of the mountains and continents, as well as the depression of the ocean bottom, make the real figure of the solid earth slightly irregular. In considering the general figure of the earth, geodesists conceive of it as if the earth had been put into a turning lathe and all the mountains and continents planed off to the sea level. The figure thus formed by the surface of ocean and planed-off land, is called the *geoid*. The figure and size of this supposed body are taken as the true figure and size of the earth, on which the continents and mountains are regarded as excrescences.

That portion of the matter composing the earth which is near its surface is called the *earth's crust*.

THE EARTH

The diameters of the geoid are: —

Equatorial diameter	7926.5 miles.
Polar diameter	7899.5 miles.

Thus the diameter of the earth is about 27 miles less through the poles than through the equator.

The surface of the geoid thus defined is everywhere at right angles to the line of gravity, which is fixed by the direction of a plumb line. Looking at figure 51, we see that this line PB does not point exactly at the center of the earth, except at the equator E and the poles. The difference between the direction of the plumb line PB and the line PC drawn to the earth's center, that is, the angle BPC, is called the *angle of the vertical*.

Its greatest amount is nearly one-fifth of a degree.

2. Latitude and Longitude. — By the *astronomical latitude* of a point on the earth's surface is meant the angle which the plumb line at that point makes with the plane of the equator. In figure 51 the latitude of the point P is the angle EBP. It is so called because it is determined by astronomical observations.

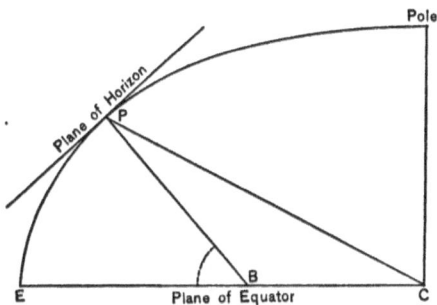

FIG. 51. — Showing the difference between geographic and geometric latitude.

The *geographical latitude* of a place is the same as its astronomical latitude, except that certain small deviations in the direction of the plumb line are allowed for.

The *geocentric latitude* is the angle which the line from the center of the earth to the place makes with the plane of the equator. In figure 51, the geocentric latitude of the point P is the angle ECP. These two latitudes differ by the angle of the vertical.

The geocentric latitude of a place cannot be directly determined, because we cannot see the center of the earth nor determine its exact direction by observation. But we can always determine the direction of the plumb line with suitable instruments. Hence on maps and for ordinary purposes the astronomical or geographic latitude is always made use of.

3. Length of a Degree. — When an observer stands on the equator, say at the point E, figure 52, the plane of his horizon is at

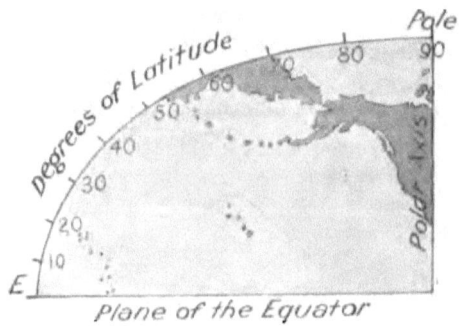

Fig. 52.

right angles to the plane of the equator. If he travels north, we say he has traveled one degree when his horizon has changed its position by one degree on the celestial sphere. The further north he goes, the slower his horizon will turn as he travels, and consequently the further he must go in order that the change may be one degree. Hence : —

The degrees of latitude are shortest at the equator, and continually grow longer as we approach the poles.

THE EARTH

They are about 68.8 miles in length at the equator and 69.3 miles at the poles.

At the equator one degree of longitude is a little more than 69 miles. Owing to the convergence of the meridians toward the poles, it continually shortens as we approach the poles, where it becomes nothing, because all the meridians there meet.

One sixtieth of a degree — that is, a minute of arc — on the earth's surface is called a *nautical mile*, because it is the mile used by sailors. The latter use it in preference to our land mile, which we call a *statute mile*, because they determine their positions by astronomical observation in degrees and minutes, and they find it easy to take one minute of arc on the earth's surface as a mile. This is nearly a mile and a sixth.

In ordinary cases navigators make no distinction between the lengths of a degree of latitude at different distances from the equator. But when we want to speak of the length of a nautical mile with exactness, we commonly take it to mean a minute of longitude at the equator.

4. How the Earth is Measured. — Owing to the obstructions on the earth's surface, and the impossibility of fixing points on the ocean, we cannot measure the distance round the earth as we would measure that round a field by a tape line. The determination of the magnitude and figure of the earth must therefore be made by special methods, in which astronomical observation and measurement of distances on the earth are combined. The operation of measuring large portions of the earth's surface with great exactness is called *geodesy*.

Geodesy requires two operations. One of these consists in determining the exact distance between two points on the earth in meters, yards, or miles. This is done by a process called *triangulation*. The other operation consists in finding out by astronomical observation what fraction of the distance round the earth, or how many degrees on its surface, is included between two points whose distance is measured by triangulation.

94   ASTRONOMY

Triangulation. — The principle on which triangulation is effected is this: The length of a line, AB, figure 53, is measured as exactly as possible on some nearly level plane.

A line thus measured for geodetic purposes is called a *base line*. The direction of the base line, or the angle which it makes with the meridian, is determined by astronomical observation.

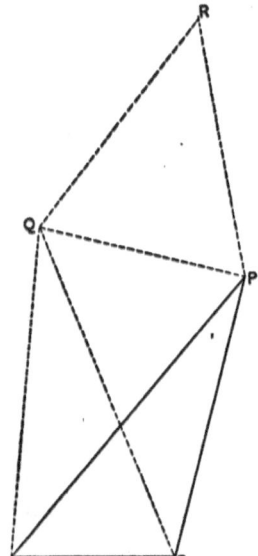

FIG. 53. — Example of a triangulation.

A distant high point P is then chosen, on a hill or mountain, which can be seen from both ends of the base line. The angles PAB and PBA are measured as exactly as possible with a theodolite. Then, by trigonometry, the sides of the triangle BP and AP can be exactly computed. If there are other distant points, like Q, which are visible from the two ends of the base line, triangles to them are determined in the same way, and their distance from each other computed.

Any of the sides AP, BP, AQ, BQ, or PQ can then be used as a new base line, and the positions of other distant points like R determined by sighting on them. By sights from these points other yet more distant points can be determined, and so on all the way across a continent if necessary. The more mountainous a region is, the easier it is to make a triangulation, because longer sights from one mountain top to another can be taken than on a plain.

Triangulation on a large scale is carried on by the United States Coast and Geodetic Survey. The latter has made measures across the American continent from the Atlantic to the Pacific Coast. Long networks of triangles have also been measured in various parts of Europe, Asia, and Africa.

THE EARTH

If the earth were a perfect sphere, the determination of its magnitude by triangulation and astronomical observation combined, would be a simple problem. Suppose we should measure a north and south arc 500 miles in length. By astronomical observation we find the difference of latitude between its two ends to be 7° 12'. Since 360° reach round the earth, we could state the proportion: —

$$7° \; 12' : 500 :: 360° : \text{circumference of earth.}$$

This would give 25,000 miles as the circumference. We might also get the length of one degree by dividing 500 by 7.2; then the circumference, by multiplying the quotient by 360. Dividing the circumference by 3.1416 would give us the earth's diameter.

This shows only the principle by which the problem is solved. The actual work is a great deal more complicated and occupies the time of many men, year after year. The complications arise not only from the ellipticity of the earth, but from the fact that wherever we go, the direction of the plumb line is slightly changed by the attraction of hills, mountains, and continents, and also by that of matter of different densities under the earth's surface. Even when these irregularities are allowed for, it is found that the figure of the geoid has many irregularities which have not yet been well determined.

5. How Latitude and Longitude are Determined. — The second operation of geodesy which we have described requires the determination of the exact latitude and longitude of places on the earth's surface. This determination is necessary, not only for the purposes of geodesy, but in order that we may make exact maps of counties and states, lay down on them the position of cities, and find the distance from one point to another. When once the size and figure of the earth are known, it is simpler to find these positions and distances by astronomical observation than it is to measure them by triangulation.

Latitude. — To determine the latitude of a place, the astronomer determines the exact point in the celestial sphere which corresponds to his zenith. He might do this in a rough way by sighting upward on a plumb line, and noticing what stars were near his zenith, but he could not get any exact result by such a process as this. The principle of the method now commonly employed is this: —

Imagine a telescope pointed nearly at the zenith, and a spirit level like that used by masons and architects, only much more sensitive, to be attached to it. Fancy the telescope to be fastened to a vertical axis which turns on a pivot at the bottom. Adjust this axis so that as we turn the telescope round, the level shall always read the same. Then we know that the line of sight of this telescope will describe a circle on the celestial sphere with the exact zenith in its center. The astronomer finds a pair of stars at the north and south points of this circle, of which he knows the declinations. Half the sum of these declinations is the declination of the zenith. This is equal to the latitude of the place, as will be seen by §§ 11 and 12 of Chapter I.

Yet other methods may be used. We have explained that the latitude of a place is equal to the altitude of the pole above the horizon. It is also equal to the angle between the zenith of the place and the celestial equator. The astronomical observer can determine these angles with great precision, by specially constructed instruments, and can thus obtain his latitude without knowing the declinations of any stars.

Longitude. — We have already shown that the difference of longitude corresponds to the difference in the local time at two places, 15° of longitude always corresponding to one hour's difference of time. We may also define the difference of time as equal to the time which it takes noon to travel from one place to the other. To show how these principles are applied, suppose that an observer at New York telegraphs to San Francisco the exact moment at which the sun crosses his meridian. Then when the sun gets to San Francisco, an observer there telegraphs to New York the moment the sun is passing the meridian of San Francisco. The elapsed time between the two signals would be the time required by noon to travel from one city to the other. Multiplying the hours, minutes, and seconds by 15 would then give us the degrees, minutes, and seconds of difference of longitude between the two cities.

The same result is found by each observer telegraphing the other at a given moment the exact sidereal time at his place. He finds the time by noting the time of transit of stars over his meridian with a transit instrument and sidereal clock. To make the determination with the greatest exactness, the clock is so arranged that its pendulum shall make a telegraphic signal which is heard at the distant station as well as recorded at the station where the clock is. Then the other observer sends a signal back from his clock, so that there are really two records of the same difference of time at the two stations. Of these differences one will be a little too great and the other a little too small in consequence of the time it takes electricity to travel from one station to the other. The mean of the two will be the correct difference of longitude in time.

A longitude thus determined is called a *telegraphic longitude*. Difference of longitude between places can be determined by skillful observers in this way with an error of only a few hundredths of a second of time, or a few yards of distance on the earth. If you should place two transit instruments three or four hundred yards east or west of each other, skillful observers would have no trouble in determining their distance apart within a few yards, by astronomical observations on the stars, combined with electric signals in the way described.

6. Density of the Earth, Gravity, etc. — By the density of the earth is meant the average specific gravity of the material composing it, or the average weight of a cubic foot of the earth's matter compared with that of a cubic foot of water. As the earth is composed of many different materials, the specific gravity of various portions of it is very different. What we want is the mean density of the whole earth.

98 ASTRONOMY

We can find out what materials compose the interior of the earth only by digging mines so as to get at them. But we cannot dig to any great depth; only in the rarest cases can we go three or four thousand feet below the surface; hence we know nothing of the materials that compose the great bulk of the interior of the earth. But we can determine the mean density of these materials by measuring the attraction of bodies whose mass is known.

Attraction of a Sphere. — Imagine a sphere of lead a yard in diameter. Since, by the law of gravitation, every particle of matter attracts every other particle, it follows that this sphere of lead must attract small bodies near it. The attraction is indeed very minute; it can be made sensible only by exceedingly delicate instruments. But in recent times methods have been contrived by which this very small force can be measured with great exactness. We have to see how, from the attraction of the sphere of lead, we can determine the mean density of the earth.

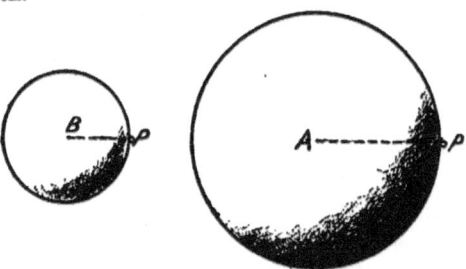

FIG. 54. — Attraction of two spheres of different sizes.

Consider two spheres of matter, A and B, figure 54, of the same density, each attracting a particle P at its surface. Let A be twice the diameter of B. It is found by mathematical processes that each sphere attracts an external body as if the whole matter of the sphere were concentrated in its center.

Sphere A being twice the diameter of B, has eight times its mass. Attraction varying directly as the mass, it will exert eight times the attraction of B at the same distance.

Attraction also varies inversely as the square of the distance, and P is twice as far from the center of A as from that of B. Hence the same matter at the center of A will attract P one fourth as much as if it were at the center of B.

There being eight times as much matter in A, its actual attraction on P will be double that of B. That is: —

Spheres of equal density attract bodies at their surfaces with a force which varies directly as their diameter.

Hence, if we find the attraction of a sphere of lead, and multiply it by the number of times the diameter of the earth exceeds that of the sphere, we shall have the attraction of a ball of lead the size of the earth. Comparing this with the attraction of the earth, we shall have the ratio between the density of the earth and that of lead. It is thus found that: —

Mean density of the earth $= 5\frac{1}{2}$ times that of water.

This is much greater than the density of the materials composing the earth's crust, and results from the enormous force with which the interior of our globe is compressed by the weight of the matter around it.

7. Condition of the Earth's Interior. — It is a very curious fact that when a mine is sunk in the earth the temperature is found to increase with the depth. The rate of increase is different in different places, but is commonly not far from 1° Fahr. in 50 feet. At the depth of 3000 feet the temperature would therefore, generally, be 60° above the mean at the surface. This temperature is so high that miners cannot live and work at the bottom of a deep mine except by having cool air pumped down to them.

There is every reason to believe that the increase continues to a great depth at the same rate, so that a few miles underground the whole earth must be red-hot. At a still greater depth, the temperature is probably sufficient to melt all the materials of which the earth is composed. This fact, taken in connection with the phenomena of volcanoes, has led to the view that the earth is really a mass of melted matter, with a hard crust a few miles thick, on which we live. But it is found that the earth does not yield to the attractive forces of the sun and moon as it would if it were liquid. Hence, the view now generally accepted is that the materials in the interior are kept solid by the enormous pressure to which the whole interior of the earth is subjected by the mutual gravitation of its parts. How great this pressure is can be conceived when we reflect that every square foot a hundred miles below the surface will be pressed by the weight of a column of earth a foot square and a hundred miles high. This weight would be not far from 40,000 tons. Such a pressure would crush any substance at the earth's surface. The reason the substance inside the earth is not crushed is that it is pressed equally on all sides so that it is merely condensed into a smaller space and made solid.

8. The Atmosphere. — The atmosphere is densest at the surface of the earth, and grows rarer as we ascend in it. This is due to the fact that every part of it is pressed by the weight of the whole mass of air above it. At the height of three miles the air becomes so rare that most people find a difficulty in breathing at such a height, and the difficulty, of course, increases with the height.

The temperature of the air continually diminishes as we ascend. Even on a summer day it is generally freezing cold at the height of a few miles. Hail is due to the freezing of raindrops in the cooler region of the air.

Air is not perfectly transparent, although it seems to be so when we look through only short distances. We all know that when we look at objects at a great distance they have a blurred, hazy aspect, due to the imperfect transparency of the air through which the light comes. The light from the heavenly bodies, as it passes through the air, is diminished before it reaches our eyes through part of it being absorbed by the air as it passes. The loss is smallest at the zenith, and increases near the horizon, because the rays of light have to pass through a greater distance in the air when the body from which they come is near the horizon.

The blue rays are more absorbed than the red rays; hence a larger proportion of red light than of blue light reaches our eyes from the heavenly bodies, and the latter look more or less red when near the horizon. This is why the sun and moon have a reddish tinge when rising or setting.

The air also reflects a small part of the light which passes through it; were it not for this, the sky would be as dark by day as by night, and we should see the stars all day. There would be no twilight, because darkness would come on as soon as the sun had set. Twilight is caused by the reflection of the sunlight from the upper part of the air after the sun has set to us.

Twilight ends when the sun is about 18° below the horizon. This shows that the air reflects no sunlight at a height greater than 45 miles. This is, therefore, commonly taken as the limit of the earth's atmosphere. But the phenomena of shooting stars, of which we shall speak in a subsequent chapter, show that there is really some kind of an atmosphere at a height of nearly 100 miles. But we do not certainly know what this atmosphere is.

9. The Zodiacal Light. — If we look at the western sky on a clear evening of winter or spring just after the end of twilight, we shall see a very faint, soft column of light extending along the region of the ecliptic, and gradually fading away as we look farther from the horizon. The same appearance may be seen in the eastern horizon before daybreak, in the summer and autumn. This appearance is called the *zodiacal light*, because it extends along the region of the zodiac. In our latitudes we cannot see it in the evenings of summer and autumn, because then the ecliptic is too near the horizon, and the light is absorbed by the thickness of the air through which it has to

102　　　　　　　　*ASTRONOMY*

pass. Within the tropics it may be seen on every clear evening.

There is something mysterious about the zodiacal light, but it is probably caused by masses of very tenuous matter, like fine particles of dust, which circulate around the sun in the whole region inside the orbit of Mars. What we see is the sunlight reflected from these very minute particles.

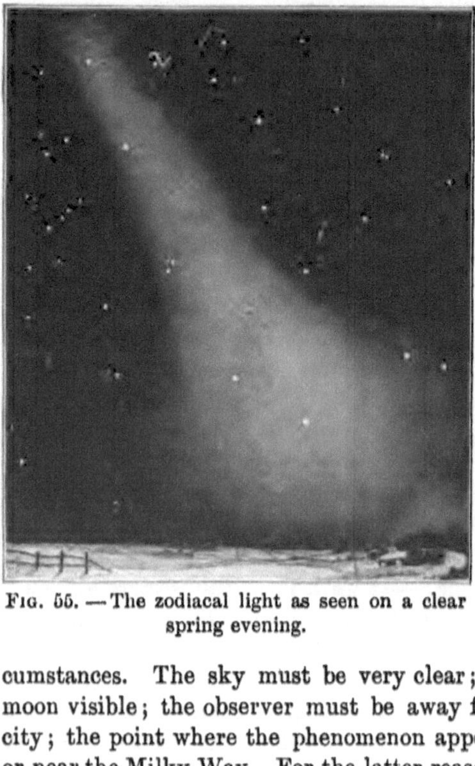

Connected with the zodiacal light is a phenomenon called the *Gegenschein*, a German word signifying *counter-glow*. It is an extremely faint light in the zodiac, exactly opposite the direction of the sun. Its faintness is such that an ordinary observer would never notice it, nor can it be seen except under the most favorable circumstances. The sky must be very clear; there must be no moon visible; the observer must be away from the lights of a city; the point where the phenomenon appears must not be in or near the Milky Way. For the latter reason the Gegenschein is not visible in June or July, nor in December or January. Even under the most favorable circumstances, the observer must have some practice in seeing a faint light in order to distinguish it. Its cause is still involved in mystery.

Fig. 55. — The zodiacal light as seen on a clear spring evening.

CHAPTER VII

THE SUN

1. Particulars about the Sun. — The sun is a globe whose diameter is more than a hundred times the diameter of the earth. Hence the distance round it is more than a hundred times that round the earth. Because the volumes of globes vary as the cubes of their diameters, it follows that the volume of the sun is more than a million times that of the earth. More exactly, it is 1,297,000 times that of the earth. You understand, without further explanation, that ·the sun looks small because of its great distance of 93,000,000 of miles, a distance which the swiftest train would not run in a hundred years.

Fig. 56. — Showing how an image of the sun may be thrown on a screen with a spyglass.

The Sun's Density, Mass, and Gravity. — The density or specific gravity of the matter composing the sun is less than that of the matter composing the earth. The mass of the sun is

104　ASTRONOMY

about 333,000 times the mass of the earth, instead of a million times and more, as it would be if it had the same density.

In consequence of its great mass, the attraction of gravitation at the surface of the sun is about 27 times the attraction of the earth on bodies at its surface. A pound of matter on the earth would weigh 27 pounds on the sun. Under an attraction so great, a man of ordinary size would weigh two or three tons, and would therefore be crushed to death by his own weight.

The Photosphere. — When astronomers speak of the sun they mean the whole body of the sun, inside as well as outside. But we cannot see the inside of the sun; we can see only its

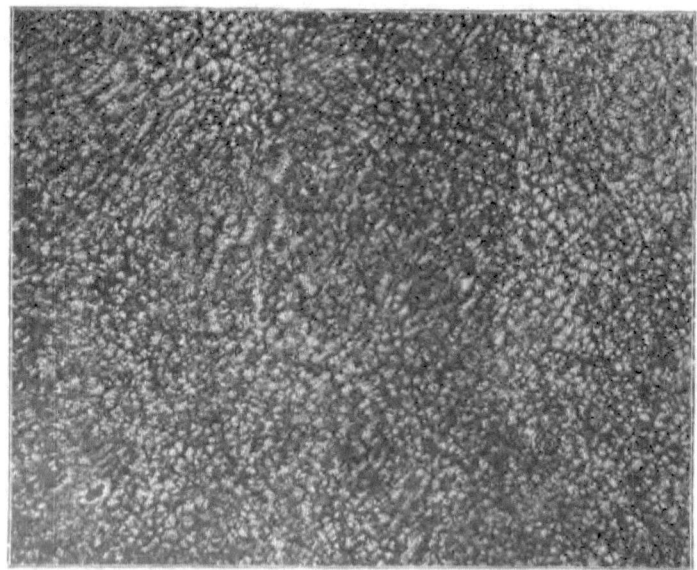

FIG. 57. — Mottling of the sun as photographed by Janssen.

surface. This visible surface is called the *photosphere* or light-sphere, because it is the part of the sun which sends us light and heat.

THE SUN

When we look at the sun with a good telescope, we see that the photosphere presents a mottled appearance, like a plate of rice soup. The grains which produce this mottling are hundreds of miles in extent. They are probably caused by the matter of the photosphere, as it cools off, continually falling back into the still hotter interior of the sun, and its place being taken by gaseous matter arising from inside, as we shall next describe.

2. Heat of the Sun. — The sun shines in consequence of its very high temperature, as iron shines when we make it red-hot. But the temperature of the photosphere is much higher than that of red-hot iron; higher than the burning coal in the hottest furnace. Possibly the temperature in the most powerful electric furnace is nearly equal to that of the photosphere. The inside of the sun is far hotter than the photosphere, and there is reason to believe that it gets hotter and hotter toward the center.

This heat is so intense that, subjected to it, all substances known to us would boil away like water over a fire and thus be transformed into vapor. Hence it is believed that matter cannot exist in a solid state in the sun. The vapor into which the substances composing the sun are changed by the fervent heat is so compressed by the enormous gravitation of the mass of the matter around it that it is forced into something between a gas and a liquid. The intense elastic force of this gaseous matter causes portions of it to be continually thrown up to the sun's surface, or to the region of the photosphere. There it speedily gets colder by radiating heat into space, and portions of it perhaps condense into solids, much as a red-hot crust will form on the surface of a pot of melted iron taken out of a furnace.

It would not be correct to say that the matter of the sun is burning, because things are said to burn when they unite with the oxygen of the air, thus producing light and heat. The sun is so much hotter than an ordinary fire that its substance could not burn. In other words, it differs from an immense fire in being so much hotter. If the earth should fall into the sun, everything on its surface would be melted in an instant, as if a small ball of wax fell into the hottest furnace.

All life on the surface of the earth is sustained by the heat of the sun, which is radiated to us as heat from a fire in an open fireplace is radiated to all parts of a room. If the sun should cease to give us heat, the air and the whole surface of the earth would slowly cool off. In a few days it would be freezing cold, even at the equator. In a few weeks the whole ocean would freeze over, and the soil would freeze to such a depth as to kill every plant. Men and animals might be able to keep alive for a while by artificial heat, but they would soon starve in consequence of not having anything to eat.

3. Spots and Rotation of the Sun. — When the sun is viewed through a telescope, dark looking spots are frequently seen on his surface. These spots are not really dark, but would seem of dazzling brightness against the sky if the rest of the sun were not there. They look dark only in contrast to the intense brightness of the photosphere.

The spots are of various sizes and shapes. Occasionally one appears so large as to be visible to the naked eye. Commonly, however, they can be seen only with the telescope. Sometimes a number of small ones are clustered together, forming a group of spots.

The spots are extremely irregular, as may be seen from the figures which we give.

The central part of a spot is the darkest. It is called the *umbra*, or *nucleus*. Around this nucleus is a border, intermediate in brightness between the darkness of the spot and the brilliancy of the photosphere. This border is called the *penumbra*.

When a spot is carefully examined with a good telescope in a steady atmosphere, it is found to be striated, looking much like the bottom of a thatched roof, the separate straws bending

THE SUN

toward the interior of the spot. This appearance is shown in figures 59 and 60.

Astronomers are not agreed as to the nature or cause of these spots on the sun, though they have been studied for nearly three centuries.

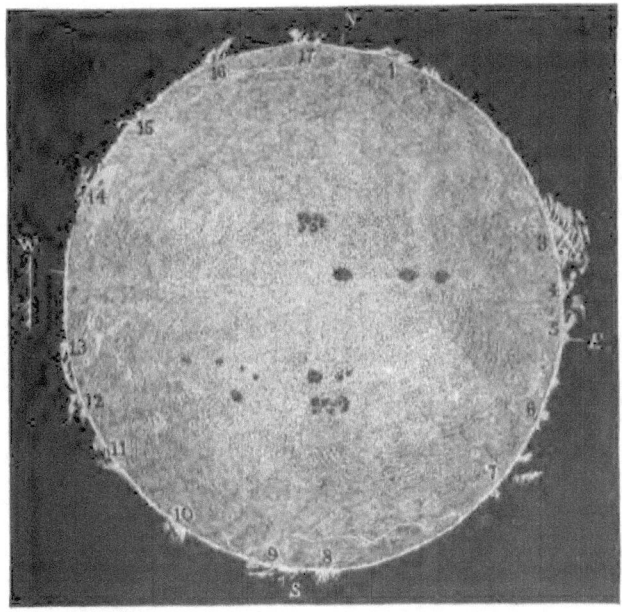

FIG. 58. — The sun, with its spots and prominences, the latter being shown by the spectroscope.

The Sun's Rotation. — When the spots are carefully watched they are seen to change their position from day to day by moving slowly across the photosphere from east toward west. In this way it is found that the sun, like the earth, rotates on an axis. The time of rotation is about 26 days. The points where the sun's axis of rotation intersect its surface are called the *poles* of the sun.

A belt around the sun, 90° from each pole, is called the sun's *equator*. It is found by watching the spots that they make a revolution in a little less time when on the equator than when at a distance from it.

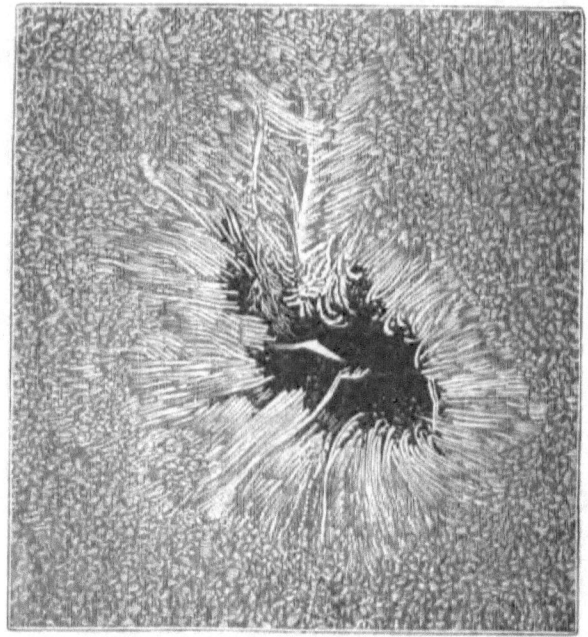

Fig. 59. — A typical solar spot, after Langley, showing the forms which such a spot often presents.

Periodicity of the Spots. — The spots are much more numerous in some years than in others. In years when spots are scarce, there will sometimes be none visible for several days, and at other times only one or two will be seen. In years when spots are numerous, quite a number of them, and sometimes very large ones, will be seen nearly all the time. It is found by the records of the years when spots were numerous

THE SUN

and when they were scarce, that there is a period of about eleven years, during one half of which there are few spots, while during the other half there are many.

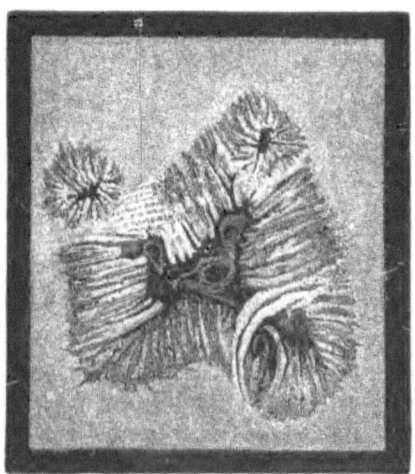

Fig. 60. — A solar spot, after Secchi.

4. Corona and Prominences. — When the sun is totally eclipsed by the moon, very curious and beautiful phenomena are seen. One of these consist of red patches or cloudlike forms around the body of the moon. These objects are called *prominences* or *protuberances*, and are found to belong to the sun. They cannot be seen with a telescope when there is no eclipse, because of the intense light of the sun dazzling our eyes. This is why we see them only when the light of the sun is cut off by the moon. But with a spectroscope they are visible on almost any clear day. This instrument shows that they are composed of masses of gas, mostly hydrogen, which are from time to time shot up from the photosphere.

Sometimes they have the form of immense flames blazing up suddenly to a height of many thousand miles with a velocity of more than 100 miles a second. In such flames everything on the surface of the earth would be destroyed in an instant.

Another object surrounding the sun is a beautiful effulgence called the sun's corona. It cannot be seen, even with a spectroscope, except during a total eclipse of the sun. It will, therefore, be described in the chapter on eclipses.

5. Source and Period of the Sun's Heat. — The sun, as we have already explained, is merely an extremely hot body radiating heat to us, as a white-hot globe of iron would radiate heat if hung in the middle of a room. One of the most interesting and important questions in astronomy is why the sun does not cool off and thus gradually cease to give us light and heat, as the iron globe would do. If, as is generally supposed, the earth and sun are many millions of years old, then the sun must have been radiating heat during this immense period. We must, therefore, account not only for the heat the sun now gives us, but for the heat which it has radiated to the earth in past ages. One explanation that has been proposed is that of meteors. It is now believed that there are great numbers of small bodies moving round the sun in its immediate neighborhood, and it is quite likely that such bodies might, from time to time, fall into the sun. Each body thus falling would generate heat. But this view is now generally given up, because it seems hardly possible that meteors in sufficient number to generate the sun's heat could be falling into the sun.

The view now commonly held is that the heat of the sun is kept up by the constant contraction of its mass through the gravitation of its particles toward the center. The theory of energy teaches us that heat is produced when a body falls toward a center without having its velocity increased. For example, the temperature of the water of Niagara Falls must be about one-quarter of a degree higher after it strikes the bottom than it is before it goes over the falls. As the sun cools off it must grow smaller, so that its outer portions fall toward its center. In this falling so much heat is acquired that, if the sun remains gaseous, it will continually grow hotter. It may, therefore, continue to radiate the same amount of heat every year, so long as it does not become a solid.

If this view be correct, a time must come when the sun can contract no more. Then a solid crust will form over its surface, this crust will gradually cool off by the heat which it radiates, and the sun will gradually grow dark and cold. But the period necessary for this is many millions of years, so we need not trouble ourselves about it.

A question of more immediate concern is whether the quantity of heat which the sun gives us is subject to variations. We know that at one time, probably not many thousand years ago, the whole of New England and the northern states was buried all the year round in snow and ice. The time when this was the case is called the *Glacial Epoch*. It is possible that during the glacial epoch the sun gave less heat than it does now. If so, it may again give less heat at some future time.

There is, however, no evidence of any change in the temperature of the earth since the invention of the thermometer. The meteorological observations made two or three hundred years ago give about the same mean temperature that they do in our time. But the earlier observations of this kind are, perhaps, not very reliable, so that their evidence cannot be conclusive against a very small change of one or two degrees. But the observations made at the Greenwich Observatory from 1840 to 1890 show that there was no perceptible change during those fifty years. This disproves a view which has sometimes been maintained, that the variations in the solar spots produce corresponding variations in the temperature of the earth. It also leads us to believe that there will be no change in the amount of the sun's heat for many years to come.

CHAPTER VIII

THE MOON AND ECLIPSES

1. Distance, Size, and Aspect of the Moon. — The moon is a globe like the earth. It looks flat to the eye because we cannot see its roundness without a telescope. In a telescope we can see it to be round like a globe.

Distance. — The moon is much nearer to us than any other of the heavenly bodies. Its average distance is a little less

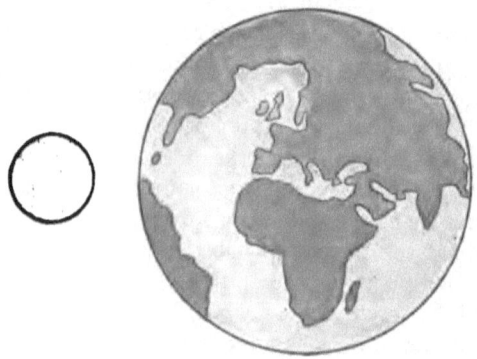

FIG. 61. — Showing the relative size of the earth and the moon. The diameter of the moon is a little more than one fourth that of the earth.

than 240,000 miles. The diameter of the earth being nearly 8000 miles, the distance of the moon is about 30 times the diameter of the earth, and therefore 60 times its radius. A railway train running 60 miles an hour would reach the moon in five or six months. An idea of the relation between the

THE MOON AND ECLIPSES 113

distances of the sun and moon may be gained by remembering that the sun is nearly 400 times as far as the moon.

Size and Density. — The diameter of the moon is about 2160 miles. This is a little more than $\frac{1}{4}$ of the diameter of the earth. In bulk it is about $\frac{1}{50}$ that of the earth. But the materials which compose it are not so dense as those of the earth. They have about 3 or 4 times the density of water. Thus the mass of the moon is about $\frac{1}{80}$ that of the earth.

Moon ---------------------- 240,000 miles ---------------------- Earth

FIG. 62. — Showing the size and distance of the earth and moon nearly in their true proportions. Their distance apart is about 30 diameters of the earth and more than 110 that of the moon.

The Moon's Surface. — If we look carefully at the moon near the time of first quarter we shall see little irregularities near the left-hand edge of the bright surface. Through a telescope this edge looks very jagged. This is because the surface of the moon has mountains and valleys upon it. Sixty or seventy of these mountains are more than a mile high, and a few are four miles or upward. They are therefore nearly as high as the highest mountains on the earth.

But the shape of the mountains on the moon is very different from that of our mountains (see figures 63 and 64). Their tops are frequently rounded like the rim of a saucer or shallow plate, the inside being hollow, and black like the bottom of the plate. In the center of this flat region there is very frequently a little sharp conical peak.

These appearances make it probable that long ages ago these mountains were volcanoes. There is, in fact, a remarkable resemblance between these lunar hollows and the craters of volcanoes like Vesuvius. A hundred years ago it was thought that there was a volcano in eruption on the moon; but we now know that this was a mistake. What was seen was only a spot of unusual brightness.

Some parts of the moon are much darker than the general surface. It is said that Galileo and others who first used a

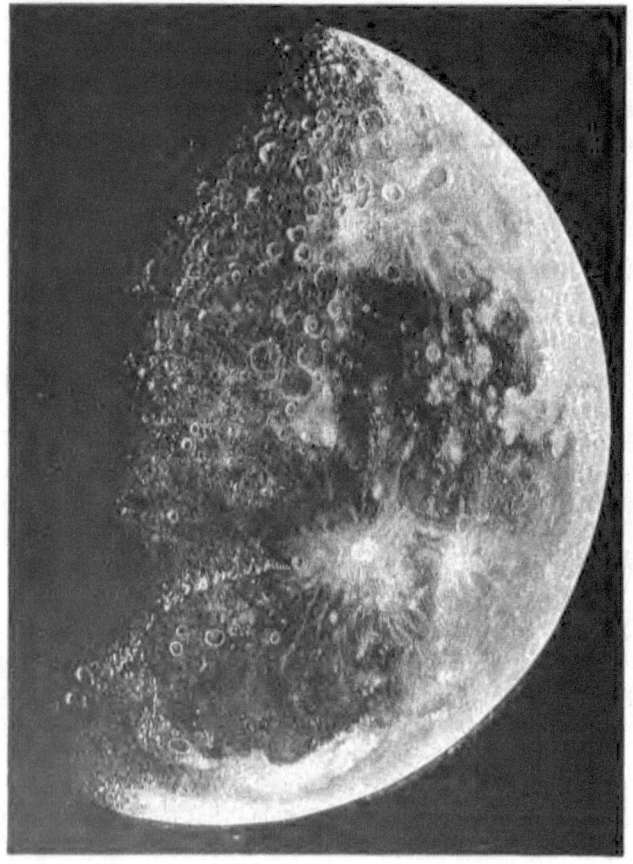

Fig. 63. — The moon, photographed by Dr. Henry Draper.

telescope supposed these dark portions to be seas, because they looked smoother than the others. Thus Milton, in allusion to

THE MOON AND ECLIPSES 115

Galileo, who was a native of Tuscany, says of Satan's shield that it

> "Hung on his shoulders like the moon whose orb
> Through optic glass the Tuscan artist views
> At evening, from the top of Fesolé
> Or in Valdarno, to descry new lands,
> Rivers or mountains in her spotty globe."

FIG. 64. — Telescopic view of a region on the moon.

But when more powerful telescopes were made, these supposed seas were found to have mountains and valleys like the rest of the surface. The darkness was merely the result of a difference of shade in the matter forming different parts of the moon.

Absence of Air and Water. — It is now certain that the moon has neither water nor air in any quantity sufficient for us to detect its existence. Consequently there is no weather on the moon and, so far as we have yet discovered, nothing ever happens there, except that the surface gets warm when the sun shines on it and cold when it does not.

2. The Moon's Revolution. — We must think of the earth and moon as two companions, revolving round the sun together, while, at the same time, they revolve round each other. The exact truth is that they both revolve round their common center of gravity, while the earth goes round the sun in the orbit we have described. Let E be the center of the earth, M that of the moon, and C their common center of gravity.

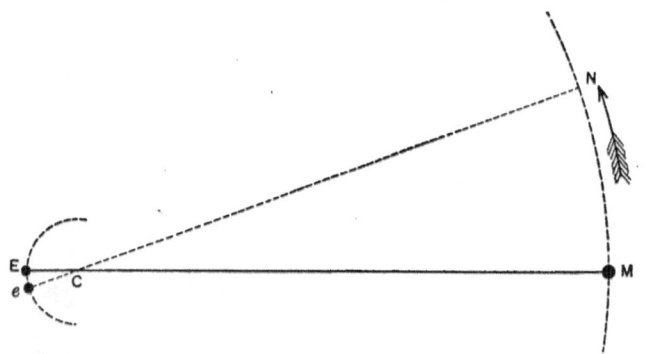

FIG. 65. — As the moon moves from M to N, the center of the earth describes the small arc Ee in the opposite direction, both moving round the common center of gravity at C.

Then EM will be the *radius vector* of the moon, which means the line from the center of the earth to that of the moon. We must now conceive that this radius vector turns round on C as on a pivot, so that, while the moon is moving from M to N, the earth moves from E to e. Thus the center of the earth describes the small dotted circle, while at the same time the moon describes the larger circle MN, of which only an arc is shown in the diagram. This combined motion arises from the fact that the moon attracts the earth as much as the earth does the moon.

THE MOON AND ECLIPSES 117

The common center of gravity C is really inside the earth, about one fourth of the way from its circumference to its center. Its distance from the earth's center is therefore so small that we commonly speak of the moon as revolving round the earth, without reference to the motion of the earth itself round C.

Sidereal and Synodic Revolution. — Let ABC be an arc of the earth's orbit round the sun. Let us start with the earth at A, and around it the orbit of the moon, with the moon at M, between the earth and the sun. In this position the moon is said to be in *conjunction* with the sun.

While the earth is moving from A to B, the moon makes one revolution around it, and reaches the point N such that the line BN is parallel to the line AM. These lines being parallel, the moon has made a complete revolution, and is seen in the same real direction at N as she was at M.

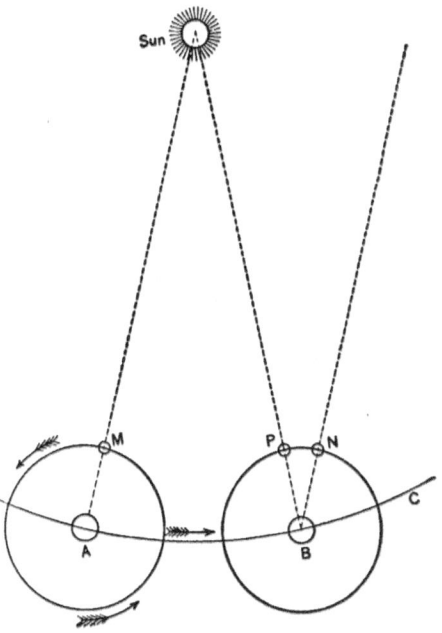

Fig. 66. — Showing the difference between the sidereal and synodic periods of the moon.

This revolution of the moon around the earth is called a *sidereal revolution* because, when it is completed, the moon has returned to the same apparent point among the stars. It takes place in about 27 d. 8 h.

Although the moon has actually made one revolution round

the earth when it comes to N, yet she will not be in conjunction with the sun at N, but will have to move through an arc NP to catch up to where the sun appears to be. This takes it more than two days more. Thus the time between the moon's conjunctions with the sun is on the average 29 d. 13 h. This period between two conjunctions with the sun is called a *synodic revolution*.

3. The Moon's Phases and Rotation. — The moon is an opaque body which shines only by reflecting the light of the sun. That hemisphere which is toward the sun is always brightly illuminated by the sun's rays; the other is in darkness so that we do not plainly see it.

When the moon is in conjunction with the sun, her dark side is turned toward us, and we cannot see her at all. The almanacs then call it *new moon*, though we cannot see the moon.

Two or three days later she has moved away from the sun so far that a small portion of her illuminated hemisphere is visible. The form which she then shows, and with which we are so familiar, is called a *crescent*, because the moon is then increasing.

At this time, if we look carefully, we shall see the entire round disk of the moon, the dark part having a very faint gray illumination. This is caused by the light from the earth being reflected upon the moon. The earth being several times larger shines much more brightly upon the moon than the moon does upon the earth. The appearance is familiarly called "the old moon in the new moon's arms."

In three or four days more the moon has got to the position of *first quarter*. One half the illuminated hemisphere is now visible to us and her visible disk has the form of a semicircle.

During the next few days we see more and more of the illuminated hemisphere, and the moon is said to be *gibbous*.

When the moon gets opposite the sun she presents the same

face to the earth and to the sun. We see her whole illuminated hemisphere and call it *full moon.*

During the second half of the revolution the phases recur in the reverse order, and a week after full moon she has got round through another quarter of her journey. We then say that she is in her *third quarter.* We can then again see one half the illuminated hemisphere.

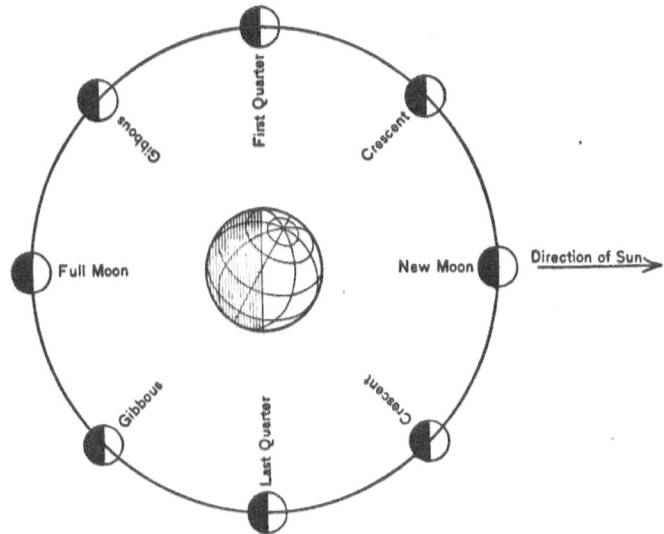

FIG. 67. — The moon's phases.

In 7 or 8 days more she is again in conjunction with the sun and we lose sight of her.

The *age of the moon* is the time elapsed since new moon. When we first see her as a thin crescent after sunset, she is commonly 2 or 3 days old. At first quarter she is 7 or 8; at full moon about 15; at last quarter 22 days old.

The best time to see the moon through a telescope is not when she is full, as people commonly suppose, but when she is between 4 and 8 days old.

Form of the Moon's Orbit. — We shall explain in another chapter that the planets move round the sun in ellipses having very nearly the form of a circle. If the earth and moon were attracted by no body but the sun, they would move around each other in ellipses, as the planets move round the sun. But the sun attracts both, and thus prevents the orbit being an exact ellipse, and also makes it change its form slightly, but continually. The result is that the orbit is much like a moving ellipse. The point of this ellipse where the moon comes nearest the earth is called the *perigee;* that where she is farthest is called the *apogee*.

The positions of the apogee and perigee are continually changing, and they make a complete revolution round the earth in about nine years.

The Moon's Effect on the Weather. — It used to be supposed that the moon had some effect on the weather, and that changes of weather were more likely to occur at new or full moon, or at one of the quarters. It is now known that this is not the case. The most careful observations show that the moon has no effect at all on the weather.

Rotation of the Moon. — As the moon revolves around the earth, she always presents nearly the same face toward us. This shows that she turns on her axis in the same time that she revolves around the earth. It should be noticed that if the moon did not turn on her axis at all, then as she went round the earth we should see her from various directions, and so should get a view of all parts of her surface.

As she always turns the same face toward us, it follows that we can never see the other side of the moon. But there are small changes in the speed with which she performs her revolution round the earth, while her rotation on her axis is uniform. Hence we can sometimes see a little farther on one side or the other of her body. Such an appearance is called *libration*. This word means a balancing, and is applied because, to our eyes, the moon seems to have a slight swing back and forth on her axis, as a balance has when the weights in the pans are equal.

THE MOON AND ECLIPSES

4. The Tides. — In consequence of its gravitation, the earth attracts the moon and thus keeps her in her orbit. If it were not for this attraction the moon would gradually leave the earth altogether, as has already been explained. But, by the third law of motion, the moon attracts the earth as well as the earth the moon. Hence the earth is being continually drawn toward the moon. But it can never move far in consequence of this drawing, because of the constantly changing direction in which the moon acts: at one time of the month the attraction is in one direction, and at the opposite time in the other direction.

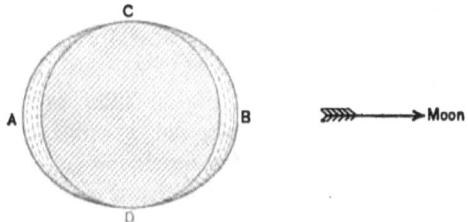

FIG. 68. — Showing how the moon causes the tides.

We have already said that gravitation is less the greater the distance. Hence the portion of the earth near the moon is attracted more strongly than the portion most distant from it. The result is that the attraction of the moon tends to draw the earth out into an ellipsoidal form. The earth itself, however, being a solid body, cannot be stretched out by this increased attraction. But the water of the ocean, being movable, is stretched out a little. Thus a wave, very broad, but only a few feet deep, is made in the ocean, and follows the moon around every day. There is also a similar wave on the opposite side of the earth. This is because at that point the water is attracted less than the average of the solid earth, so that the moon pulls the earth away from the water. Thus there are two waves a day moving round the earth.

These waves are called *tidal waves*. The rise and fall of the water of the ocean which they produce are called *tides*. They strike our coast and make the water rise for 6 hours, until the top of the wave reaches us. It is then called *high tide*. During the next 6 hours the tide recedes. At its lowest it is called *low tide*. Six hours later there is another high tide, and so on. Thus there is a regular rise and fall of the water twice every day, with which all who live on the seacoast are familiar.

In consequence of the continual motion of the moon on the celestial sphere, from west toward east, she passes the meridian on the average about 50 minutes later every day than she did the day before. Hence, the tides arrive later every day by this average amount.

The amount of the rise and fall is very different in different regions. Out in the ocean it is generally less than on the coast, commonly only 2 or 3 feet. As the tidal wave approaches a coast the resistance of the latter causes the water to pile itself up against the coast, and thus rise to a height of 6, 10, or 20 feet, or, in rare cases, much more.

Owing to the islands and continents, the tidal wave is not merely one wave going along uniformly, but sometimes there are several waves in different parts of the same ocean. When two of these waves happen to meet, they make one big wave. If there happens to be a deep, wide-mouthed bay where they meet, the water may rise to a very great height in consequence of the force with which it enters the bay. This is the case in the Bay of Fundy, on the coast of Nova Scotia and New Brunswick. At the head of this bay the tides rise 70 or 80 feet. The effect is here most extraordinary. The Basin of Minas is quite a large lake, at high tide being 12 miles across and 40 miles long. But at low tide it is almost empty.

Spring and Neap Tides. — The attraction of the sun on the earth produces a tide as well as that of the moon. But this tide is smaller than that of the moon. At the times of new and full moon, the sun and moon unite their attraction to produce tides. Consequently the tides are higher at those times than at others. These are called *spring tides*.

At first and last quarter the sun and moon pull against each other on the tides. Thus the sun diminishes the effect of the moon, and the tides are not so high. They are then called *neap tides.*

Another effect of this combined action of the sun and moon is that the actual intervals between the high tides on successive days sometimes vary considerably from the average interval. Sometimes high or low tide occurs at nearly the same time on two successive days. At other times the difference of time may be more than an hour.

5. Eclipses of the Moon. — All opaque bodies cast shadows when the sun shines on them. Hence the moon and the earth cast shadows. Night is caused by our being in the shadow of the earth when our hemisphere is turned away from the sun.

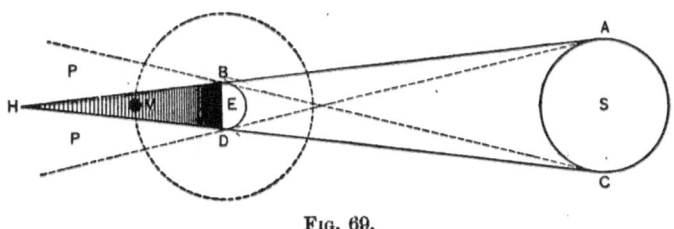

FIG. 69.

Let S, figure 69, be the sun, E the earth, and M the moon. Draw the lines ABH and CDH meeting at H, and touching the sun and earth. You will then see that between these two lines, in the region between the earth and H, the light of the sun will be cut off. This region is that of the *shadow* of the earth. The shadow has the shape of a cone, with its point at H. This is called the *shadow cone.*

Outside the shadow is a region PP in which the light of the sun is partly but not wholly cut off. An observer in this region, if he could fly up to a great distance from the earth, would see the latter hide a greater or less part of the sun, according to his nearness to the surface of the shadow cone.

The region PP in which the sunlight is partly, but not wholly, cut off, is called the *penumbra.*

When the moon is entirely in the shadow of the earth, the direct light of the sun can no longer reach her, so she looks dark. We then say that there is an *eclipse of the moon*. That is, an eclipse of the moon is caused by the moon passing through the shadow of the earth.

It is very interesting to watch such an eclipse. As the moon enters the shadow we see a small part of one edge of her disk grow dark and finally disappear. The darkness spreads over the disk little by little until it covers the whole surface of the moon. During the first part of the eclipse we cannot see the eclipsed portion of the moon because of the dazzling effect of

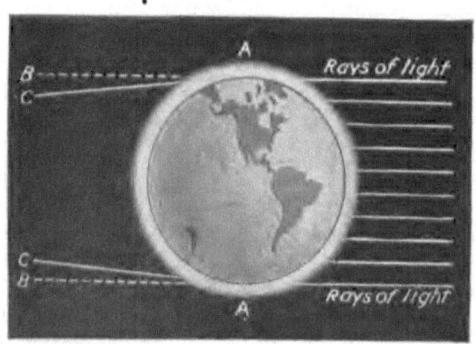

Fig. 70. — Refraction of the light of the sun into the earth's shadow.

the bright part. But when the bright part has nearly or quite disappeared, we see the whole disk shining with a dim, reddish light. This is because the light of the sun is refracted by the earth's atmosphere as we have explained in Chapter IV, and shown by the above figure. Hence the rays of the sun, which pass very near the surface of the earth, are so refracted by the air that they enter the shadow, and keep it from being perfectly dark. To an observer on the moon, looking at the earth during an eclipse, the sun would be entirely hidden by the earth, but the latter would be surrounded by a thin ring of this refracted light, of a reddish tint. This tint is due to the absorption of

THE MOON AND ECLIPSES

the blue rays by the atmosphere, and hence arises from the same cause that makes the sun look red when on the horizon.

Sometimes only a part of the moon dips into the shadow. The eclipse is then called a *partial eclipse* of the moon.

When the moon is altogether immersed in the earth's shadow, the eclipse is said to be *total*.

6. The Moon's Orbit and Nodes. — You may now ask why it is that there is not an eclipse of the moon at every full moon, because the moon is then always opposite the sun. The reason is that the shadow of the earth is always in the ecliptic, while the orbit of the moon around the earth is inclined to the plane of the ecliptic, as shown in figure 71.

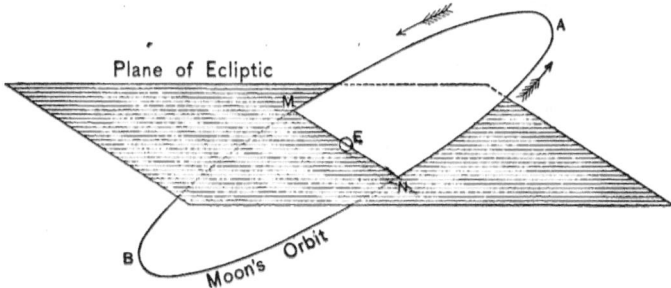

FIG. 71. — Showing the inclination of the moon's orbit to the plane of the ecliptic. The latter is represented by the horizontal surface; the plane of the moon's orbit by the inclined circle. The latter intersects the ecliptic at the points *M* and *N*, which are called nodes. The line joining the two nodes is called the line of nodes.

If we imagine ourselves standing on the earth at *E*, and mapping out the moon's course among the stars, as we have imagined the apparent course of the sun to be mapped out, the two courses would not be the same, but would intersect each other at two opposite points. When the moon was in the half *A* of her orbit, she would appear north of the plane of the ecliptic, and when in the half *B*, south of it.

There are two opposite points, *M* and *N*, at which the orbit

126　　　　　　　　*ASTRONOMY*

intersects the ecliptic. These points are called *nodes*. The straight line joining the nodes passes through the center of the earth, where also the plane of the moon's orbit intersects that of the ecliptic. It is called the *line of the nodes*.

Figure 72 shows four positions of the earth's shadow, when it happens to be near one of the moon's nodes. As the moon moves along her orbit she crosses the ecliptic, and, if the shadow happens to be near the same point, may enter it in whole or in part, according to the distance from the node.

Fig. 72. — The dotted circles show different positions of the earth's shadow near the moon's node. The shadow is always opposite the sun; but we cannot really see it. As the moon passes along it may enter the shadow partially or entirely, then we see it more or less eclipsed. The smaller circles represent the moon. In the two right hand positions the moon is wholly immersed in the shadow; in the left hand position it does not enter the shadow at all.

The inclination of the moon's orbit to the ecliptic is about 5°. This is ten times the apparent diameter of the moon.

7. Eclipses of the Sun. — An *eclipse of the sun* is a partial or entire hiding of the sun through the intervention of the moon.

Of course the moon casts a shadow as the earth does. Figure 73 shows its form. We draw the lines *SM* and *TN* from the edge of the sun, meeting in the point *H*. Here the shadow ends in a sharp point. To an observer in the shadow the sun will be completely hidden by the dark body of the moon. The eclipse is then said to be *total*.

Now draw lines *TM* and *SN*, touching the moon. Between these lines and the shadow is the penumbra. An observer in this region will see the sun partly hidden by the moon. This phenomenon is called a *partial eclipse* of the sun.

THE MOON AND ECLIPSES 127

The average distances of the sun and moon from the earth are such that the point *H* of the shadow is generally very near the surface of the earth. When the moon is near perigee the shadow does not quite come to a point before reaching the earth. There will then be a small region of the earth's surface

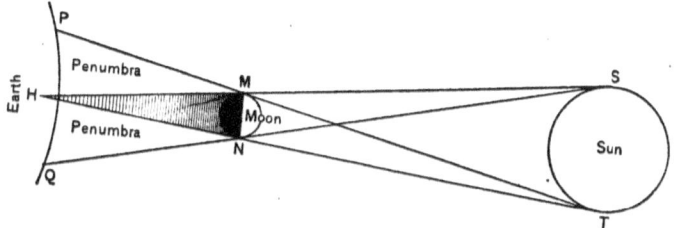

FIG. 73. — The moon's shadow and penumbra. This figure shows the shape of the dark shadow of the moon. To an observer inside the shadow the moon will entirely hide the sun. An observer in the penumbra will see the sun partially covered by the moon. An observer at the point of the shadow will see the moon exactly cover the sun. Sometimes the surface of the earth is just at the point, sometimes inside and sometimes outside, according to the varying distance of the moon.

on which the eclipse will be total. As the moon moves in its orbit the shadow sweeps along a path on the earth's surface. This is called the *path of total eclipse*. An observer cannot see the eclipse as total unless he places himself somewhere along this path. In the astronomical ephemeris maps are given showing the paths of all the total eclipses that occur. These are of various breadths, but are generally less than a hundred miles wide.

There can be no eclipse unless, at the time of new moon, the sun is near one of the moon's nodes. If this is not the case the moon will seem to pass above or below the sun. The various kinds of eclipses of the sun can be best understood by studying figure 74. The dark body of the moon is shown in five positions, while it is passing the sun, as it would appear if we could see it, which, however, we cannot do except as we may see the sun eclipsed.

In position *A*, when the sun has not reached the node, the moon passes a little below the sun. This causes a partial eclipse, the appearance of which the figure shows.

In the position *B*, the centers of the sun and moon coincide exactly at the moment of conjunction. The eclipse is then said to be *central*. There are two kinds of central eclipses: total and annular.

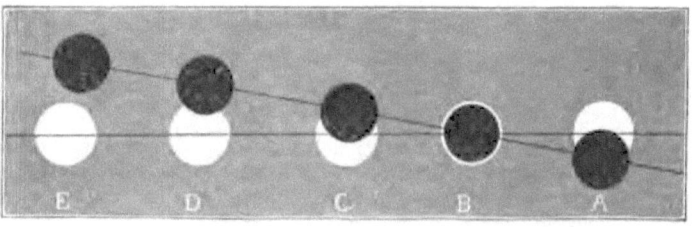

FIG. 74. — This figure shows how at new moon there may be a central, total or annular eclipse of the sun, a partial eclipse, or no eclipse at all, according to the distance of the sun from the moon's node. In each figure the round white circle represents the sun and the black circle the dark invisible body of the moon which may pass over it. In the position *B* the sun is exactly at the moon's node, so that the dark body of the moon passes centrally over the sun. At *A* the moon passes a little below the sun and at *C* a little above it, covering the greater part. When the sun is still farther from the node, as at *D*, the moon covers only a small portion of it. When the sun is still farther, as at *E*, the moon passes above it entirely so that there is no eclipse. When the moon is off the sun, as at *E*, we cannot see it in the heavens but we can imagine its position as shown in the figure.

If the apparent diameter of the moon is greater than that of the sun, as will be the case when the moon is near perigee, the sun will be entirely hidden and the eclipse will be *total*.

If the apparent diameter of the moon is a little less than that of the sun, which will be the case when the moon is near apogee, the edge of the sun's disk will be seen all around the disk of the moon. The eclipse is then called *annular*, because the edge of the sun is seen as a ring (*annulus*, a ring).

If the moon passes the sun in the position C or D, there will be a large or a small partial eclipse, as the moon passes.

If the sun is far from the node, as at E, the moon will pass clear of the sun and there will be no eclipse at all.

As the moon's shadow and penumbra pass over the earth, the diameter of the penumbra is a little more than half that of the earth. Hence the sun will never be eclipsed all over the earth, but only in those parts over which the penumbra or shadow sweeps. If an observer in one part of the earth saw a central eclipse, like B, an observer farther south would see the moon pass north of the sun's center, as in C, D, E, and there would be a large partial eclipse, a small one, or no eclipse at all, according to his position.

Of course every eclipse of the sun begins by being small and partial, and gradually increases as the opaque body of the moon advances. The various appearances of the eclipse as it advances are called *phases* of the eclipse.

A total eclipse of the sun is a very impressive sight, especially if one observes it from a high elevation where he can see many miles around. Owing to the direction of the moon's motion in her orbit the shadow of the moon sweeps along the earth in an easterly direction; it may be due east, or it may incline to the north or south in a greater or less degree. During the partial phase of such an eclipse the observer will see nothing very striking except the gradual covering up of the sun's disk, reducing it to the form of a crescent. When it is almost covered, if he looks in the direction from which the shadow is coming, he will see the darkness approaching, perhaps at the rate of nearly a mile a second. As the shadow reaches him the sun entirely disappears. Looking up, he now sees the black body of the moon with the sun's corona around it. The latter is a most beautiful effulgence, like the glory which the old painters depicted round the heads of their saints. It is quite irregular in shape, parts of it extending out in streamers. It shades off so gradually that no distinct outline can be seen. When viewed with a telescope the corona is seen to have a somewhat woolly or fibrous structure, which is also shown on the photographs.

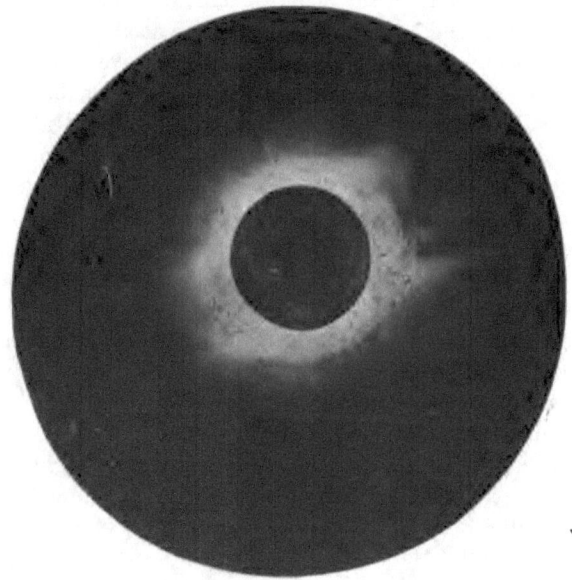

FIG. 75. — The solar corona during a total eclipse of the sun.

The light of the corona probably comes from very minute vaporous particles shot up from the sun, and perhaps held up by some form of magnetic or electric action. Part of the light may also come from clouds of particles revolving round the sun. The brightest stars will also be visible. It will not, however, be entirely dark, but, as described by Milton, the sun

> ". . . From behind the Moon
> . . . disastrous twilight sheds."

This is because the sun is shining through the air all around the region of total eclipse, and the light reflected from without

THE MOON AND ECLIPSES 131

this region penetrates the whole shadow, and enables us to see surrounding objects. The darkness is about that which we have half an hour after sunset. The time by a watch can be seen during the whole of the eclipse.

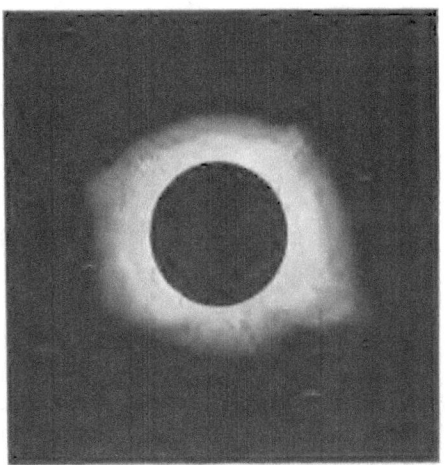

Fig. 76. — Another view of the solar corona.

8. Recurrence of Eclipses. — Since there are two opposite nodes of the moon's orbit, and the sun makes an apparent circuit of the heavens in the course of a year, it follows that the sun will appear to pass the moon's nodes twice in every year, at an interval of about six months. Hence eclipses of the sun or moon may occur at two opposite times of the year, about six months apart. We may call these times eclipse seasons.

As an eclipse may occur at either node of the moon's orbit, it frequently happens that, if there is an eclipse of the moon at one node, then, when she makes half a revolution, which takes about fifteen days, there will be an eclipse of the sun at the opposite node, and *vice versa*.

If the position of the moon's nodes were invariable, the eclipse seasons would always be the same, year after year.

132   ASTRONOMY

But the position of the node is continually changing, owing to the attraction of the sun on the earth and moon. The manner in which this change takes place will be seen by studying figure 77. Here the small circles show ten different positions of the moon as she is passing her node, the position of her orbit being shown by the dotted line *mm* through the centers.

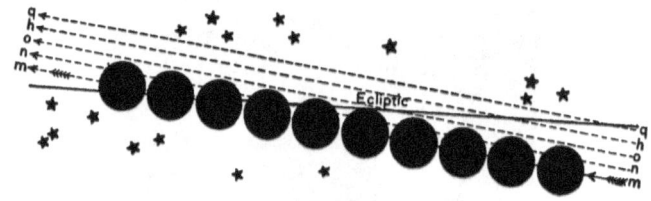

Fig. 77.

In the next to the last position she is exactly at the node, and is therefore crossing the ecliptic. But when she makes a revolution and gets back to the same position in the heavens, she will not follow exactly the same path, but will follow along the line *nn*. In another revolution she will pass along the line *oo*, and so on continually. At the fifth revolution the point of crossing or node will be nearly at the right hand end of the figure. Hence the node is in motion from east toward west. In this way each node makes a complete revolution in the heavens in eighteen years and about seven months. The result of this is that the eclipse seasons occur about twenty days earlier every year than they did the year before, because the sun in its apparent path catches up to the node that much sooner every year.

CHAPTER IX

THE CALENDAR

A *calendar* is a system of defining, arranging, and numbering days, months, and years, so as to form a continuous measure of time to be used by the people of a country. In former ages, when people were not in so close communication as they are now, each nation generally had its own calendar. But at the present time nearly all civilized nations have adopted the one used by us.

1. Units of Time. — The first and most natural unit of time to be adopted by men is the *day;* understanding by that term the period between two successive passages of the sun over the meridian, which, as we have already explained, is slightly greater than the true time of one revolution of the earth on its axis.[1] The use of this unit of time was enforced upon men from the beginning by the alternation of the activity of the day with the repose of the night.

The next period of time to be noted was the *year*. The cycle of the seasons, which the earliest men who noticed the order of nature must have seen to be due to the varying declination of the sun, determines the year. One of the earliest works of men who made astronomical observations was to determine the exact times at which the sun reached the equinoxes in

[1] It is an unfortunate defect of our language, as of most modern languages, that the word *day* is used in two senses : (1) the period during which the sun is above the horizon, in contradistinction to *night* when the sun is below the horizon ; and (2) the length of a day and night together. The reader will, in each case, see for himself in which sense the term is used.

successive years. The number of days between two returns to the same equinox defined the length of the year. Very early in history it was thus discovered that the length of the year was about $365\frac{1}{4}$ days.

The Month and Easter. — So large a number as this was inconvenient to keep count of. An intermediate unit was therefore necessary. This was afforded by the changes of the moon. At intervals, which our ancestors found to be about thirty days, the moon completed the circuit of the heavens, disappeared in the sun's rays, and again reappeared in the west after sunset. The reappearing body was called the new moon. The interval between two successive new moons is called the *lunar month*.

In very ancient times, as we know from the Bible and other writings, the moon was used to determine the times of certain religious festivals. This practice survives with us in the date of Easter, which is determined as follows: —

The first full moon after the 21st of March in every year is called the *Paschal full moon*. *Easter Sunday* is the Sunday which follows this full moon. If the latter occurs on Sunday, Easter is the Sunday following.

The moon goes through its cycle of changes a little more than twelve times in a year. Had it done so exactly twelve times, there would have been no difficulty in using it as a measure of months. This not being the case, it was impossible to make a true year out of 12 lunar months. A year thus measured was only $354\frac{1}{2}$ days in length. Those who used it found the seasons in the course of 30 years to wander through every part of the year, which was inconvenient.

Another method was to fit the lunar months and years together, a year having sometimes 12 months and sometimes 13. This also was inconvenient.

A third method, and the one which is now adopted by the leading nations, is to reject the lunar months altogether, and divide every year into 12 months, without any respect to the moon.

THE CALENDAR 135

2. The Julian Calendar. — Two thousand years ago the Romans were the leading nation of the world; but their calendar was in great confusion before the time of Julius Cæsar, because the emperors or other rulers fixed it from time to time to suit themselves. To remedy this confusion Julius Cæsar arranged what is now known after him as the *Julian Calendar*. He (or his learned men) saw that if we had in regular succession three years of 365 days and then one year of 366 days, the average length of the year would be 365¼ days, which was then supposed to be the true length. So this plan was adopted and was carried by the Romans into the various countries which they conquered. Owing to its approach to correctness it was continued after being once fully accepted. Thus it became the calendar of civilized Europe.

3. The Gregorian Calendar. — In the sixteenth century it was found that the period of 365¼ days was a little more than the true length of a year. The error was not very great; it amounted to only one day in about 130 years. Still, it had the effect, in the course of centuries, of making the equinox fall at a different time of the year from that at which it had been arranged to fall by the festivals of the Church. In the sixteenth century the error had amounted to ten days, it being assumed that the correct arrangement was that made by the Council of Nice, 325 A.D. To remedy this, Pope Gregory XIII, in 1582, modified the Julian calendar by arranging that the closing years of the centuries 1600, 1700, etc., should not be leap years unless the number of the century was divisible by four. That is to say, 1600, 2000, 2400, etc., were to be leap years as in the Julian calendar, but 1700, 1800, 1900, 2100, etc., were to be common years.

This is called the *Gregorian Calendar*, after the name of the pope who established it. It is now in use by all Christian nations except Russia and Greece. Even these are expected to adopt it sometime.

In establishing the calendar Gregory added ten to the count

of days, so as to make the years begin at the same time they would have begun had his calendar been adopted by the Council of Nice. To bring this about it was ordered that the day which, in the Julian calendar, would have been October 5, 1582, should be called October 15, so that the 15th day of that particular month followed the 4th day. This made a difference of 10 days between the Gregorian and Julian calendars, which became 11 days on March 1, 1700, because in the Julian calendar February of that year had 29 days, whereas it only had 28 in the Gregorian calendar. In 1800 the difference became 12 days, and from 1900 to 2100 it will be 13 days.

Thus, in our calendar, the rule for leap year is that every year the last two figures of whose number is divisible by 4 is a leap year, except the terminal years of a century ending in 00. These are leap years when, and only when, the number of the century is divisible by 4.

4. The Year. — The reckoning by the Julian calendar is sometimes called *Old Style*, and that by the Gregorian *New Style*. Besides the length and arrangement of the year, a calendar must determine two things: at what time of the period of 365 or 366 days a new year shall begin; and from what epoch the years shall be counted. Even after the adoption of the Julian and Gregorian calendars, there was some difference among nations as to the beginning of the year. In England it commenced March 1 instead of January 1. The change to the latter date was made in 1752, and is now universal.

In ancient times it was the custom to count years from the accession of some monarch, or the foundation of the government, or of its capital city. Thus the Romans counted their years from the supposed foundation of the city of Rome. We see a survival of this custom among us when, in official documents, the year of the Independence of the United States is given. But for the purposes of civilized life, our practice of counting the years from the birth of Christ has become coextensive with the Julian and Gregorian calendars.

THE CALENDAR 137

5. Features of the Church Calendar. — In our religious festivals there still survive some remains of the ancient attempts to arrange the measure of time by the moon. One of these is the *Metonic Cycle*, called after Meton, a Greek who lived about 433 B.C. He found that a period or cycle of 6940 days could be divided up into 235 lunar months, or 19 solar years.

In consequence of 19 years being nearly an exact number of lunar months, Easter will commonly, though not universally, fall upon the same day after a period of 19 years. Hence, if we count the years from 1 to 19, and then begin over again, the dates of Easter will be repeated in regular order. This number, which ranges from 1 to 19, is called the *Golden number*. It is said to owe its name to the enthusiasm of the Greeks over Meton's discovery, who caused the division and numbering of the years on the plan of Meton to be inscribed on public monuments in letters of gold. The rule for finding the golden number is to divide the number of the year by 19 and add 1 to the remainder. It is employed for finding the time of Easter Sunday.

In our Church calendars a system is used for indicating the day of the week on which any given date will fall. January 1 is marked by the letter *A*, January 2 by *B*, and so on to *G*, when the letters begin over again, and are repeated through the year in the same order. Thus each letter in any one year indicates the same day of the week through the year, except in leap years, when February 29 and March 1 are marked by the same letter, so that a change occurs at the beginning of March. The letter corresponding to Sunday of any year is called the *Dominical*, or *Sunday letter* for that year. When we once know what letter it is, all the Sundays of the year are indicated by it. In leap years there are two dominical letters, one extending to the end of February, and the other through the remaining ten months of the year.

We shall find by making the calculation that 28 Julian years contain exactly 1461 weeks: It follows that at the end of 28 years the dominical letter will be repeated as before. This period of 28 years is called the *solar cycle.*

Note that

$$28 \times 365\tfrac{1}{4} = 10227 = 7 \times 1461.$$

An exception, however, occurs when we pass over a centennial year which is not a leap year, as in 1900. For example, the dominical letter will not be the same in 1909 as it was in 1890, 19 years before. But the solar cycle will go on regularly through the twentieth and twenty-first centuries, a break again occurring in the year 2100.

6. The Hours. — In ancient times the period from sunrise to sunset was divided into 12 hours, which were counted from 1 to 12. Thus men spoke of the first hour of the day, meaning the first hour after sunrise, the second, etc., a system quite familiar to readers of the Scriptures. The third hour was that when the middle of the forenoon was approaching; the sixth hour terminated at noon; the ninth hour terminated at the middle of the afternoon, and the twelfth hour at sunset. The night was also divided in the same way into 12 hours: the first hour of the night, the second, etc., to the twelfth.

The day being longer in summer than in winter, the length of the hours thus defined changed with the seasons, and the night hours were shortest when the day hours were longest. As people had no clocks or watches in those days, and no exact instruments for measuring time were kept in the house, this variability of the length of the hours did not cause any serious inconvenience. If the hours had been of equal length, this system of counting 12 hours of the day and then 12 hours of the night would have been, in some respects, more convenient than our own, because the count of hours would have gone on continuously through the day, and again through the night. But the practice of beginning a new day at sunset was found to be inconvenient, because our activities always continue into the night; and confusion would arise from passing from one day to another at 6 o'clock in the evening.

THE CALENDAR

When the day was taken to begin at midnight, it was in some places the practice to count the hours from 1 to 24. Then during the forenoon the count of hours would have been the same as that we use up to noon. But what we call 1 o'clock would have been 13 o'clock, and so on to 24 o'clock, which would have occurred at midnight.

Probably men found it inconvenient to carry on a count of hours greater than 12. In consequence the practice was introduced of beginning the count of hours over again at noon, as we do, and indicating which of the 12-hour periods was meant by the words A.M. and P.M., abbreviations of the terms *ante meridiem* and *post meridiem*. This beginning of a new count of 12 hours at noon is frequently an inconvenience. Hence efforts are being made in some quarters to induce people to count the hours up to 24. On the Italian and some Canadian railways this is done in the time-tables, and it would avoid much trouble if it were done everywhere, and if we all counted the hours up to 24 from one midnight to the next.

Astronomers count the hours from 0 to 24 on the system we have just suggested, only their day begins at noon instead of midnight. This is because they use the day solely as a measure of time, without regard to light or darkness, and it is just as convenient to begin it at one moment as at another. Its beginning being determined by the passage of the mean sun across the meridian, it is more natural to count the hours from that moment. Thus, in astronomical reckoning, mean noon is 0 hour. What we call 1 o'clock P.M. is 1 hour, and so on to midnight, which is 12 hours. Then 1 o'clock in the morning is 13 hours, and so on to 11 o'clock A.M., which is 23 hours.

This mode of beginning the day and counting the hours is called *astronomical time*, while the count on the ordinary plan is called *civil time*.

CHAPTER X

GENERAL PLAN OF THE SOLAR SYSTEM.

1. Orbits of the Planets. — We explained in the second chapter that the principal bodies of the solar system are the sun and a number of planets revolving around it, on one of which we dwell. We have now to learn some general facts about the arrangement and motions of the planets.

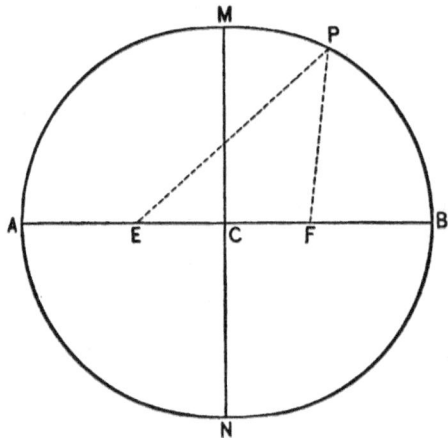

FIG. 78. — Showing how an ellipse may be drawn.

The orbits of the planets, including that of the earth, are not exact circles, but ellipses, which, however, differ so little from circles that the eye could not see the deviation.

An ellipse may be described by attaching the ends of a string to two fixed points, E and F, whose distance apart is less than the length of the string. Then by passing a pencil around the

GENERAL PLAN OF THE SOLAR SYSTEM 141

string, keeping the latter tight, the point of the pencil will describe an ellipse. Each of the points E and F, around which the ellipse is described, is called a *focus*. The center C of the ellipse is half way between the foci. The longest diameter, AB, is called the *major axis;* the shortest, MN, the *minor axis*.

The distance FC or CE between the center and each focus was formerly called the *eccentricity* of the ellipse. In modern times we call the *eccentricity* the quotient of this distance divided by the major axis. It is commonly expressed as a decimal fraction.

2. Kepler's Laws. — The laws of motion of the planets round the sun are called *Kepler's laws*, after the astronomer Kepler who discovered them.

The *radius vector* of a planet is the line from the sun to the planet. As the planet moves round the sun the radius vector is conceived to turn round the sun with it, as on a pivot.

The *mean distance* of a planet from the sun is half the sum of its greatest and least distances. It is equal to one half the major axis of the orbit.

The *periodic time* of a planet is the time it takes to make a revolution round the sun.

Kepler's laws are these: —

I. The orbit of each planet is an ellipse, having the sun in one focus.

II. As the planet moves round the sun the radius vector sweeps over equal areas in equal times.

III. The squares of the periodic times of the planets are proportional to the cubes of their mean distances from the sun.

To understand the second law, suppose figure 79 to be the orbit of a planet having the sun in the focus. Mark on the orbit the points P, Q, R, etc., which the planet reaches at equally distant intervals of time, — it may be intervals of a day, a month, or a year. Draw the radii vectors from the sun to each point. Then the triangular areas *PSQ*, *QSR*, etc., described by the radius vector will all be equal to each other.

It follows from this law that the nearer the planet is to the sun the faster it moves. We have already explained this by showing that, as the planet falls toward the sun it gains velocity, and as it recedes from the sun it loses velocity.

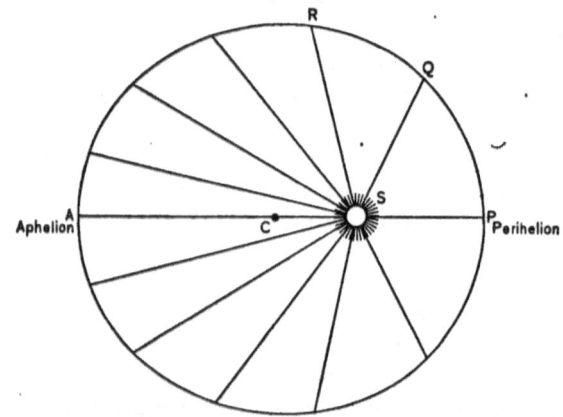

FIG. 79. — Showing how the radius vector of a planet, as the latter moves round the sun, sweeps over equal areas in equal times.

If we look at figures 78 and 79 we shall see that because the sun is not in the center of an orbit, but in one focus, there are in every orbit two opposite points at the ends of the major axis, at one of which the planet is nearest the sun, and the other farthest away.

The point of the orbit where the planet comes nearest the sun is called the *perihelion;* the point where it is farthest away is called the *aphelion*.

3. Structure of the Solar System. — The eight principal planets of the solar system are called *major planets* to distinguish them from an immense group of smaller ones called *minor planets*. All the planets, except the two nearest the sun, have one or more moons or satellites revolving round them. Thus we have

GENERAL PLAN OF THE SOLAR SYSTEM 143

to consider, not merely planets revolving round the sun, but systems each composed of a planet and its satellites, fashioned somewhat after the solar system. As the sun is the center

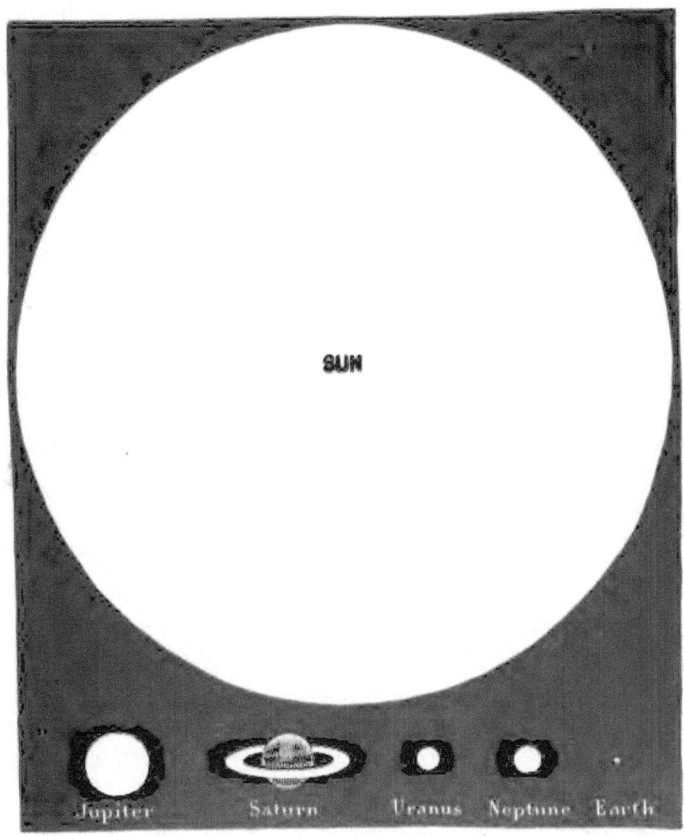

FIG. 80. — Showing the relative size of the sun and the principal planets.

around which the planets revolve, so are some of the planets centers round which satellites revolve. The satellites are generally much smaller than their central planets. Our moon is a satellite of the earth, revolving round it in the manner we have already explained.

A planet having a satellite moving round it is called a *primary planet* to distinguish it from the satellites, which are also called *secondary planets*.

The time in which a planet completes its revolution round the sun, or a satellite round its primary planet, is called its *periodic time*, or its *period*.

In addition to the major planets and their satellites, there is a curious group of bodies called *minor planets* or *asteroids* occupying a region where we might suppose there ought to be a single planet.

The following is a list of the principal bodies or groups of bodies of the solar system, with the periods of revolution of the planets round the sun:—

1. The Sun, the great central body.
2. The planet Mercury . . . . . Period 88 days.
3. The planet Venus . . . . . Period 225 days.
4. The planet Earth, with 1 satellite . . Period 1 year.
5. The planet Mars, with 2 satellites . . Period 2 years.
6. A group of several hundred minor planets, or asteroids, periods mostly from . 3 to 6 years.
7. The planet Jupiter, with 5 satellites . Period 12 years.
8. The planet Saturn, with 8 satellites . Period 29 years.
9. The planet Uranus, with 4 satellites . Period 84 years.
10. The planet Neptune, with 1 satellite . Period 165 years.

The planets Mercury and Venus, which revolve inside the orbit of the earth, are called *inferior planets*.

Those whose orbits are outside that of the earth are called *superior planets*.

4. Distances of the Planets; Bode's Law. — The distances of the planets from the sun range from 36 million miles in the case of Mercury, to 2775 million in the case of Neptune. The distance of the earth is 93 million miles. But astronomers do not use miles in celestial measurement, not only because they are too short, but because distances in miles cannot be always

GENERAL PLAN OF THE SOLAR SYSTEM 145

known exactly and other units of measurement are more convenient. To express distances of the planets they take the distance of the earth from the sun as the unit of measurement. We may call this unit a *sun-distance*. Expressed in terms of sun-distances, the distance of Mercury is 0.387, that of the Earth 1, and that of Neptune 30.

Bode's Law. — About a hundred years ago it was found by the astronomer Bode that the distances of the planets then known could be found approximately in the following way: —

Form the row of numbers 0, 3, 6, 12, each one after the second being found by multiplying the preceding one by 2. Then add 4 to each number, and divide the sum by 10, thus: —

```
        0    3    6    12   24   48   96   192   384
        4    4    4    4    4    4    4    4     4
        -    -    -    -    -    -    -    -     -
        4    7    10   16   28   52   100  196   388
÷ 10   0.4  0.7  1.0  1.6  2.8  5.2  10.0 19.6  38.8
```

These numbers may now be compared with actual distances of the planets from the sun, as follows: —

	True Dist.	Bode's Law	Error of Law	Period in Years
Mercury	0.387	0.4	.013	0.2408
Venus	0.723	0.7	.023	0.6152
Earth	1.000	1.0	.000	1.000
Mars	1.524	1.6	.076	1.881
Blank	—	2.8	—	—
Jupiter	5.203	5.2	.003	11.86
Saturn	9.539	10.0	.461	29.46
Uranus	19.18	19.6	.42	83.74
Neptune	30.04	38.8	8.76	164.78

Except for the planet Neptune, which was not known when Bode lived, the numbers found by his rule hit very near the truth, as we see by the errors given in the third column. The most interesting feature of the table is that Bode had to fill a

NEWCOMB'S ASTRON. — 10

wide gap between Mars and Jupiter, where his rule would place a planet, but where, in his time, no planet was known to exist. He therefore predicted that a planet might some time be found in this gap. Not only one, but hundreds, have been found since his time.

In 1846, when Neptune was discovered, it was seen that its distance did not agree with the rule. This shows that what is called Bode's law is not really a law of nature, but only an accidental coincidence.

We have added the more exact periodic times to the above table that the student may test Kepler's third law. Find the cubes of the several distances and the squares of the periods and compare the one with the other.

5. Aspects of the Planets. — In consequence of our being carried around the sun upon the earth, the apparent motions of the planets are different from their real motions. The varying positions of the planets relative to each other and to the sun are called *aspects*. To explain these various aspects, and show their cause, certain terms are used which we shall now define.

The *elongation* of a planet is its apparent angular distance from the sun. The word is generally applied to the elongation of Mercury or Venus. In figure 81 let the earth be in the position shown, and let the circle represent the orbit of Mercury or Venus around the sun. Let P be any position of the planet in its orbit. Then the angle between the lines ES drawn from the earth to the sun, and EP drawn from the earth to the planet, is the elongation of the planet.

The elongation is greatest when the planet is at one of the points M or N. At N it will appear east of the sun, and is then said to be at its greatest east elongation. In this case it will be visible after sunset.

When the planet is at N, it is said to be at its greatest west elongation. It may then be seen in the morning before sunrise.

GENERAL PLAN OF THE SOLAR SYSTEM 147

When two heavenly bodies, in their courses around the celestial sphere, pass by each other, they are said to be *in conjunction*.

When they are on opposite sides of the celestial sphere, so that, for example, one would be rising while the other was setting, they are said to be *in opposition*.

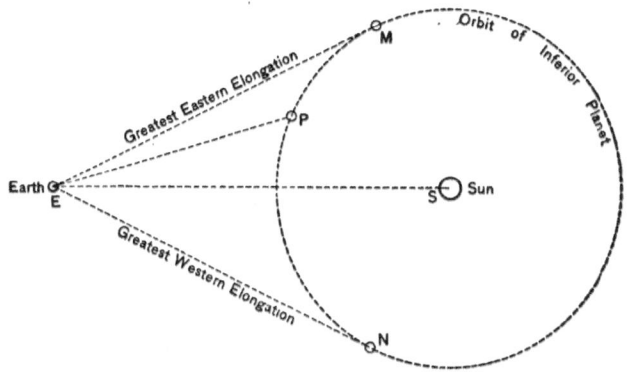

Fig. 81. — Showing the different directions in which an inferior planet may be seen from the earth. The greatest elongation is the angle between the lines so marked and the line drawn from the earth to the sun.

The conjunction of a planet with the sun is called *inferior*, when the planet passes between the sun and the earth. It is called *superior* when the planet is beyond the sun.

It is evident that the superior planets can never be in inferior conjunction with the sun because they can never pass between the earth and the sun. The inferior planets Mercury and Venus may be either in inferior or superior conjunction.

6. Apparent Motions of the Planets. — If we could view the stars and planets from the sun, each star would always appear fixed in the same place, and the planets would be seen always moving forward in their orbits, completing their revolution in the times we have mentioned. But when we consider the apparent motions, as we see them from the earth, we find that

148 ASTRONOMY

they are affected not only by these real motions, but by an apparent swing caused by our being carried around the sun upon the earth.

After making a long slow sweep toward the east for several months, or perhaps a year, the planet will gradually stop and make a short sweep toward the west. Then it will gradually stop, and again start on a long sweep eastward, and so on.

The sweep of a planet from west toward east is called *direct motion*.

The sweep from east toward west is called *retrograde motion*.

Between the east and west sweeps the planet seems nearly at rest, it is then said to be *stationary*.

Apparent Motion of a Superior Planet. — To show how the motion of the earth causes an apparent retrograde motion of a planet, let EF in figure 82 represent the orbit of the earth, and

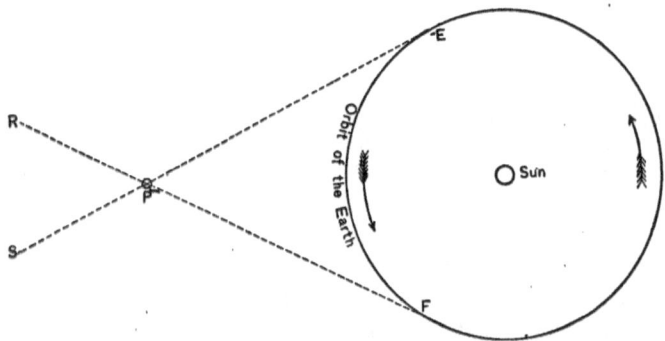

Fig. 82. — Showing the different directions in which a superior planet may be seen in consequence of the motion of the earth. As the earth swings along from E toward F it moves faster than the planet so that in the middle of the swing the planet appears to have a motion in the opposite direction.

the arrows the direction of its motion around the sun. Suppose a body to be at rest at the point P. Then, when the earth is at E, the body will be seen on the celestial sphere in the direction ES. When the earth reaches the point F the body will

GENERAL PLAN OF THE SOLAR SYSTEM   149

be seen on the celestial sphere in the direction FR. That is to say, while the earth has moved from E to F, the body, though really at rest, has seemed to swing on the celestial sphere through an arc equal to the angle EPF. This apparent motion, being the opposite of that of the earth, will be retrograde.

But if P is not a fixed body, but a planet, it is really in motion in the same direction as the earth. If it moved as fast as the earth, or faster, there would be no retrograde swing at all. But a superior planet moves more slowly than the earth. Hence there is some retrograde motion, though less than there would be were the planet at rest.

As the earth is moving through the right hand arc of its orbit, from F to E, the direct motion of the planet as we see it is increased by the effect of the earth's motion, so as to appear greater than it really is. Hence the planet makes a long direct swing.

We may, therefore, sum up the matter by saying that there is a real direct motion of each superior planet around the celestial sphere corresponding to its motion around the sun, and that the apparent deviations from this real motion are in the nature of swings due to the revolution of the earth on which we live around the sun.

Apparent Motions of the Inferior Planets. — These planets have direct and retrograde swings like the superior ones, but for reasons a little different. If such a planet were alongside the sun, the effect of the annual revolution of the earth would be that the sun would seem to us to carry the planet with it in its apparent annual motion round the celestial sphere. Hence, the effect of the earth's motion is to make the inferior planets complete an apparent revolution round the sphere in a year.

Because these planets revolve around the sun they seem to us to swing first to one side of the sun and then to the other. Hence they have apparent direct and retrograde motions, as in the case of the superior planets. During the retrograde swing the planet seems to pass the sun from east toward west; during the direct swing it passes from the west to the east of the sun.

150   ASTRONOMY

7. Perturbation of the Planets. — If a planet were attracted by no other body than the sun, it is found that it would move around the sun in an ellipse in exact accordance with Kepler's laws. This ellipse would remain in the same position forever.

But, according to the law of universal gravitation, each planet is attracted, not only by the sun, but by all other planets. In consequence of this attraction the motion of each planet deviates from the fixed orbit that it would describe if the sun alone attracted it. These deviations are called *perturbations*.

In consequence of the perturbations, each planet is sometimes a little ahead of the place it would otherwise occupy and sometimes a little behind it; sometimes a little outside the orbit and sometimes a little inside. The average orbit which the planet describes is also subject to slow changes which are called *secular variations*. This term is applied because these variations continue through many ages. Thus the eccentricity of the earth has been diminishing for many thousand years and will continue to diminish for many thousand years to come. In fact the eccentricities of the orbits of all planets are changing in this way, some increasing and others diminishing. The position of the perihelia of the earth and all the planets is slowly changing. The most rapid changes occur in the case of the moon. It is owing to the attraction of the sun that the line of nodes makes a revolution in 18 years, and that the perigee moves round in 9 years.

Ever since the time of Newton many of the ablest mathematicians in the world have investigated the laws according to which these deviations in the motions of the planets take place. The results they have reached form some of the most remarkable triumphs of the human intellect.

One result is that, by means of tables of motions of the planets, the astronomer can compute these motions for many centuries past or future, with such exactness that if the computed planet were placed along side the real one, the keenest eye could not distinguish between the two.

CHAPTER XI

THE INNER GROUP OF PLANETS

IF we examine the list of the eight major planets in § 3 of the preceding chapter we shall see that they may be divided into two groups. One group comprises the four inner ones, Mercury, Venus, the Earth, and Mars; the other, the four outer ones, Jupiter, Saturn, Uranus, and Neptune. The groups are separated by the region of minor planets. The outer group is distinguished by the great size as well as the great distance of the planets that compose it. In this chapter we shall describe the inner group, and that of the asteroids.

1. The Planet Mercury. — Mercury, the nearest planet to the sun, is much the smallest of all the major planets. Its revolution being completed in 88 days, it makes more than four revolutions a year.

Synodic Revolution of Mercury. — Suppose that at a certain time, the earth is at E, figure 83, and Mercury at M. The latter is then in inferior conjunction. At the end of 88 days it will have made a complete revolution and got back to the point M. But it will not then be in inferior conjunction, because the earth will have moved forward in its orbit. When it catches up to the earth, so as to be again in inferior conjunction, it will be at N and the earth at F. The planet is then said to have made a *synodic revolution*. The synodic period of Mercury is 116 days, or a little less than 4 months.

When Mercury is near inferior conjunction, it is invisible in consequence of the brightness of the sun's rays. A few days later it will have passed around so far as to be visible with a telescope west of the sun. A month later it will be near its greatest western elongation, and can then be seen with the naked eye, in the east, before sunrise. In another month it will be on the other side of the sun in superior conjunction, and again invisible. A month later it will be visible in the west, after sunset.

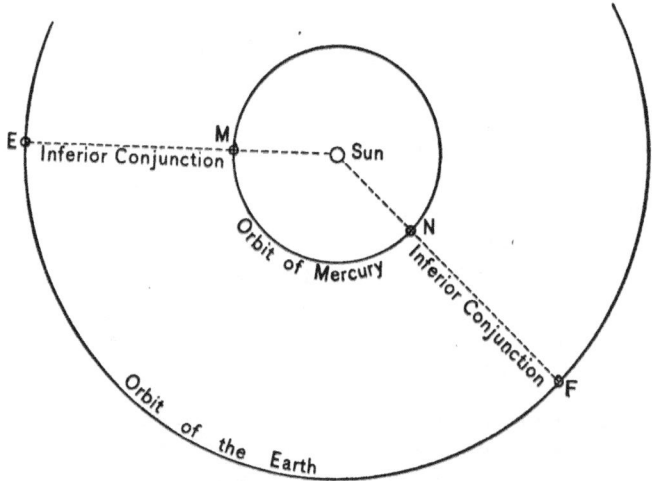

Fig. 83. — Showing the synodic period of Mercury.

Visibility of Mercury. — Near greatest elongation Mercury is very plainly visible to the naked eye, as it is then brighter than any of the fixed stars. If, looking in the west after sunset, you see a star near the horizon in the red glow of twilight, you may suppose it very likely the planet Mercury. But, to be sure, you must know the position of Venus and of the brighter fixed stars. If the object can neither be Venus nor a fixed star, it is Mercury.

Nodes and Transits of Mercury. — If the orbit of Mercury were in the plane of the ecliptic, it would pass between the the sun and the earth at every inferior conjunction. We should then see the planet as a small round spot on the sun's disk, invisible to the naked eye, but easily observed with a telescope.

But as a matter of fact, the orbit of the planet is inclined 7° to the ecliptic; hence, at nearly every inferior conjunction the planet passes below the sun, or above it. Occasionally, however, Mercury really passes between the sun and the earth. This phenomenon is called a *transit of Mercury*.

The plane of the orbit of Mercury, like that of the orbit of the moon, intersects the plane of the ecliptic at two opposite nodes. The node at which the planet passes north of the ecliptic is called the *ascending node*, that at which it passes to the south of it the *descending node*. It passes each node once in every revolution.

The line through the sun from one node to the other is called the *line of nodes*. This line intersects the orbit of the earth at two opposite points, through one of which the earth passes about May 8 of every year, and through the other about November 10. Hence when Mercury happens to pass the ascending node within two or three days of November 10, the sun, planet, and earth will be nearly in a straight line, and we see a transit of Mercury. The same things happen when it chances to pass the descending node within two or three days of May 8. The following table shows the dates of the transits between 1890 and 1950.

1891, May 9	1924, May 7
1894, November 10	1927, November 10
1907, November 14	1940, November 11
1914, November 7	

2. The Planet Venus. Aspects of Venus. — Venus is nearly the size of the earth, its diameter being only about 300 miles less than that of our globe. It performs a revolution in its orbit around the sun in 225 days, or about seven months and a half. As in the case of Mercury, it is sometimes in inferior conjunction with the sun, sometimes at western elongation; then in superior conjunction and then in eastern elongation. The synodic period is 1.6 years or about 1 y. 7 m. 3 d. This is the interval between inferior conjunctions. Five times this period is 8 years. Hence Venus has 5 synodic periods in 8 years.

At its greatest elongation Venus is about 45° from the sun. It is then the most brilliant object in the heavens except the sun and moon, sometimes casting a faint but visible shadow. Hence there is no difficulty in recognizing it when it is near either of its elongations. When at its greatest brilliancy, it can be clearly seen by a good eye in the daytime, provided that one knows exactly where to look for it.

Morning and Evening Star. — Near greatest eastern elongation, Venus, being visible in the west after sunset, is known as the evening star. The ancients then called it Hesperus. When west of the sun, it is seen in the east before sunrise, and is called the morning star. The ancients then called it Phosphorus. It is said that in the early ages people did not know that Hesperus and Phosphorus were the same body. But it required only careful comparison of observations to make it plain that this was the case.

Phases of Venus. — To the naked eye Venus always appears like a star, but with a telescope we find it to show phases like the moon. Figure 84 shows these phases. In the position A, after it has passed inferior conjunction, most of its dark hemisphere is turned toward us. But we see a small portion of the illuminated hemisphere, which gives it the apparent form of a crescent, like that of the moon when two or three days old. A few days later it has reached B. Now it is farther from the earth, so that it would look smaller, but the crescent is growing broader because we can see a larger portion of the illuminated hemisphere.

At C the planet will look like a half moon. In the following positions it will look gibbous, or round, and will appear smaller and smaller, until it reaches superior conjunction The disk will then be small and round. Completing the revolution, the same phases will recur in reverse order.

THE INNER GROUP OF PLANETS 155

Venus was one of the first objects at which Galileo pointed his telescope. He was greatly delighted when he found it to exhibit phases like those of the moon, because this showed that the planet was an opaque body revolving round the sun.

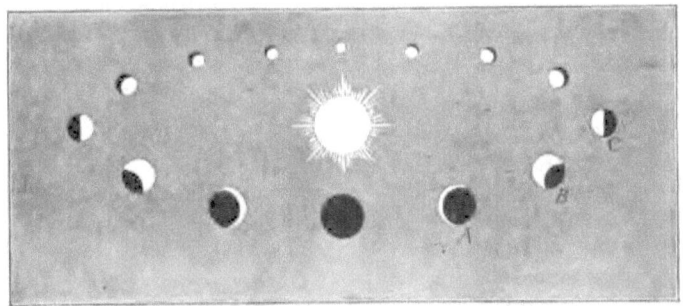

FIG. 84. — Showing the different phases of Venus during its synodic revolution.

Supposed Rotation of Venus. — When viewed with a good telescope, in a very steady atmosphere, Venus has a burnished look, something like that of polished silver shining by the light of a fire. Its disk seems brighter in the central portion, near the convex edge, than elsewhere. Some astronomers think they see very faint markings, as shown in the frontispiece. It is hence supposed by some that the planet rotates on its axis in the same time that it revolves around the sun, and thus always has the same hemisphere turned to the sun. Others have supposed that it rotates on its axis in about 24 hours, or in nearly the same time as the earth. It is still uncertain whether either view is correct.

Transits of Venus. — The orbit of Venus is inclined to the ecliptic about $3\frac{1}{2}°$. The line of nodes of its orbit cuts the earth's orbit at points through which the earth passes about June 6 and December 6 of each year. If Venus happens to pass the corresponding node within a day or two of either of these dates, we shall see her passing over the sun's disk. This phenomenon is called a *Transit of Venus*. Such transits are, however, among the rarest phenomena of astronomy. Two may occur with an interval of eight years between them. Then more than a century will elapse before there will be another. Not more than two ever occur in a century, and the whole twentieth century will pass without any at all. The following table shows the dates of all occurring between the years 1600 and 2100: —

	INTERVAL
1631, December 7	8 years
1639, December 4	121½ years
1761, June 5	8 years
1769, June 3	105½ years
1874, December 9	8 years
1882, December 6	121½ years
2004, June 8	8 years
2012, June 6	

These transits were formerly looked forward to with great interest. They were supposed to afford the best means of determining the distance of the sun, because of the considerable parallax of Venus at this time. Hence, at each of the last four transits, expeditions were sent to various parts of the world to make the most exact observations possible upon the times of the transit or the apparent position of Venus on the sun's disk. Such expeditions were sent out by the United States and other nations in 1874 and 1882 to various points in Asia, Africa and Australia. It is now found, however, that there are other methods of determining the sun's distance, more exact than this.

3. The Planet Mars. Aspects of Mars. — Mars is the fourth planet in the order of distance from the sun, and the next outside the orbit of the earth. Its mean distance from the sun is about 141 millions of miles. The eccentricity of its orbit is such that at perihelion it is only 128 millions of miles from the sun, while in aphelion it is 154 millions of miles. Hence, if at its opposition to the sun it should happen to be in perihelion, it would only be 35 millions of miles from us, the earth

THE INNER GROUP OF PLANETS 157

being 93 millions and Mars 128 millions of miles from the sun. This distance is greater than the least distance of Venus from the earth. But when the latter planet is nearest to the earth, its dark hemisphere is turned toward us, so we cannot get a view of it. But when Mars is nearest to us, its bright hemisphere is toward us, and therefore we can study it with our telescopes.

Mars comes into opposition to the sun at intervals of 2 years and 1 or 2 months. At such times it rises near the time of sunset, and may then be easily recognized by its reddish light and great brilliancy. It is, indeed, not nearly so bright as Venus, but it is brighter than any of the fixed stars. Owing to its different distances from the sun, it is much brighter at some oppositions than at others. When the opposition occurs in the month of August or September, Mars will be at its brightest; it will be faintest at the oppositions occurring in February or March.

Surface of Mars. — When Mars is examined with a powerful telescope, dark and bright regions are seen on his disk. It has frequently been supposed that the bright regions are continents and the dark regions oceans. It was found by Schiaparelli that the supposed oceans are sometimes joined together by dark streaks, which he, supposing them to be water, called channels. Schiaparelli was an Italian, and the Italian word *canale*, which means channel as well as canal, has been translated *canal*. This gave rise to the notion of canals in Mars, which we frequently read about, but which have no real existence.

Ever since astronomers began to study the supposed continents and oceans on Mars, it has been thought that this planet has a surface much like that of our earth. The resemblance is made still more remarkable by the polar caps of Mars. To understand these caps, we must know that the equator of Mars is inclined to its orbit by a rather greater angle than the equator of our earth is inclined to the ecliptic. Hence, Mars has seasons even more marked than the earth. During about one half of its revolution its north pole is turned away from the sun. During this time it is found that a bright, white cap is formed around the pole. When the sun begins to shine on the pole this cap gradually melts away, or at least becomes much

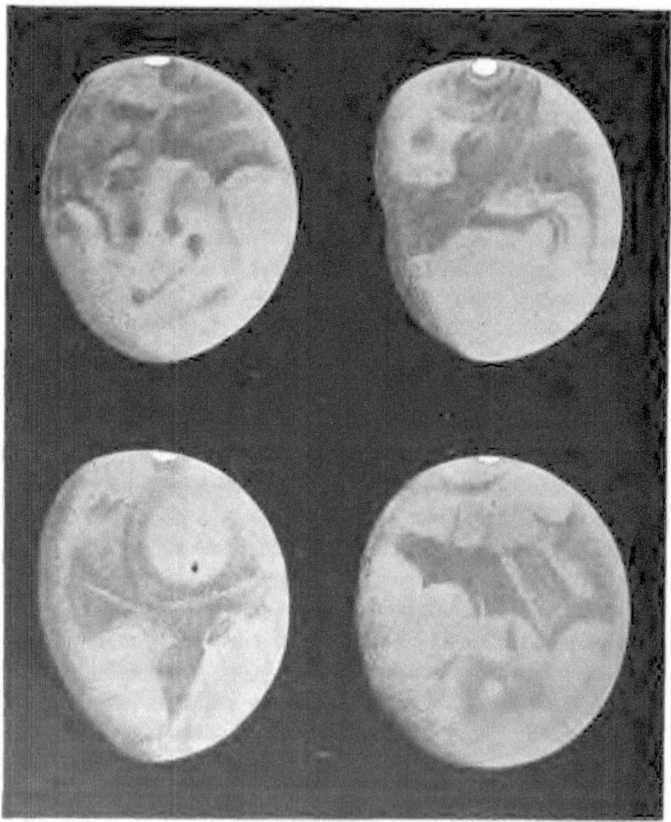

FIG. 85. — Four drawings of Mars made by Barnard with the great Lick Telescope.

smaller. The same thing takes place round the south pole of the planet during the other half of the revolution. It is therefore supposed that these caps consist of snow or frost which

THE INNER GROUP OF PLANETS 159

falls or is deposited during the Martian winter, and melts away again during the Martian summer.

The atmosphere of Mars is very rare; indeed, it is not certain that this planet has any atmosphere at all that we can get evidence of.

Rotation. — Mars revolves on its axis in 24 h. 37 m. Hence its day is a little longer than ours is.

Supposed Inhabitants. — It is generally supposed that Mars may be inhabited. This, however, is purely a supposition, because we can, with our telescopes, get no evidence on one side or the other of the question. The only reason we have to believe in such inhabitants is the fact that our earth is inhabited.

The Satellites of Mars. — These satellites were discovered by Professor Hall in August, 1877. Previous to that time it was supposed that Mars had no satellites. These bodies are the smallest yet known in the solar system, and can be seen only

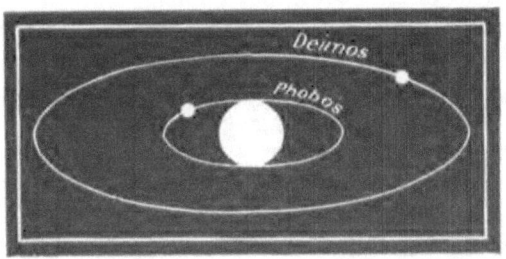

FIG. 86. — Apparent orbits of the satellites of Mars in 1877.

with powerful telescopes, and under favorable conditions. The inner one moves round its primary in a shorter time than any other known satellite, making a revolution in 7 h. 38 m. This is less than one third the day of Mars, consequently to an inhabitant of that planet Phobos rises in the west and sets in the east. Its distance from the surface of the planet is only about 4000 miles, and therefore a little more than half the diameter of the earth. Our moon is about sixty times this distance from the earth.

4. The Minor Planets or Asteroids.

Four of these planets were discovered during the early part of the nineteenth century, between the years 1801 and 1807. Then no more were found until 1845. In that year a fifth was discovered, soon after a sixth, and then they began to be found in great numbers, sometimes 12 or more in a single year. The number known is now approximating 500, and new ones are found so frequently that we do not know what the total number may be.

In recent years discoveries of these bodies have mostly been made by photography. To apply this method, photographs of the sky, the plate being exposed for several hours, are taken. The stars appear on the plates as points. But if a planet is photographed it will appear as a little line in consequence of its motion. In this way a new minor planet is found from time to time.

These bodies are far smaller than the major planets. Their size is not exactly known, but the largest are probably about 400 or 500 miles in diameter; the smallest not more than 20. The diameters of the smaller ones cannot be given with certainty, because they are so small that the planet appears only as a point of light in the largest telescope.

The eccentricities of the orbits of the minor planets are generally larger than those of other planets, and their orbits are frequently more inclined to the ecliptic. They are therefore scattered widely through the region which they occupy. Yet it is probable that if they were all combined into a single planet, the mass of this planet would be smaller than that of any major planet, even Mercury.

Olbers's Hypothesis. — When only four of these planets were known, it was supposed that they were fragments of a large planet which had, in some way, been broken to pieces. This idea was first propounded by the astronomer Olbers, hence it has since borne his name. But it is now found that this could not have been the case, because the orbits cover too much space, and have never intersected each other, as they would have done had such a breaking up occurred.

THE INNER GROUP OF PLANETS

The Planet Eros. — The most remarkable of these planets is one discovered in 1898, and called Eros. Its orbit, instead of being wholly outside that of Mars, is partly outside and partly inside. If it were not for the great inclination of the orbit of Eros, it would intersect the orbit of Mars. But it is inclined so much that it passes above the orbit of Mars at one point, and below it at another. In fact, the two orbits are linked together in such a way that, if each were made of wire, they would pass through each other like two links of a chain.

Eros, at certain times, comes nearer the earth than any other body of the solar system, the moon excepted. Hence, observations upon it may enable its distance to be measured with greater exactness than can be attained in the case of any other planet.

CHAPTER XII

THE FOUR OUTER PLANETS

1. The Planet Jupiter. — Jupiter is sometimes called the giant planet. Not only is he the largest and most massive of the planets, but his mass and size are greater than those of all the other planets combined. His mean diameter is about 85,000 miles, and his volume is therefore 1300 times that of our earth.

Jupiter rotates on his axis in the remarkably short period of 9 h., 55 m. The centrifugal force generated by this rapid rotation makes him assume a very oblate figure. His equatorial diameter exceeds his polar diameter by 5000 miles. The ellipticity of his disk is therefore plainly visible with a telescope.

A very remarkable feature of the material composing this planet is that its specific gravity is scarcely one fourth that of the earth, and only about one third greater than that of water, hence, although 1300 times the volume of the earth, the planet has only 313 times the mass of the earth.

Jupiter revolves around the sun in nearly 12 years. He comes into opposition to the sun at intervals of about 13 months. When in opposition he rises at sunset and may be well seen in the evening. In brightness he is, when seen by the naked eye, between Sirius and Venus. He can be distinguished from Mars by his whiter light, which is very different from the reddish light of Mars.

Surface of Jupiter. — The most remarkable features of the surface of Jupiter consist of certain dark, cloudlike bands, two of which, north and south of his equator, are especially

THE FOUR OUTER PLANETS 163

marked. These bands can be seen with a very small telescope, and have been known for more than two centuries.

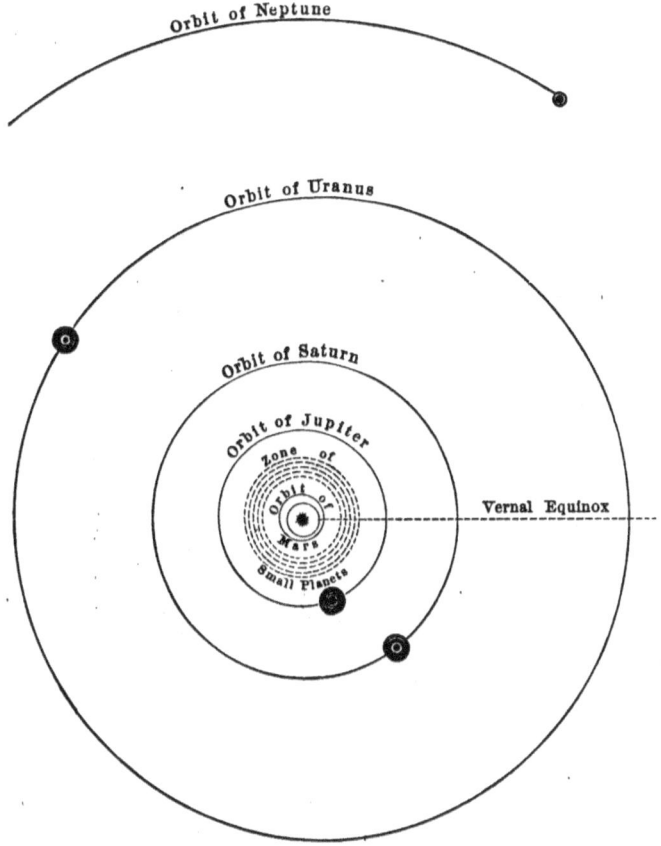

Fig. 87. — Orbits of the superior planets.

When examined with a powerful telescope it is seen that not only these bands, but the entire surface of the planet, are variegated with shadings, differing greatly in form and brilliancy. The bands especially consist of great numbers of stratified, cloudlike appearances. Sometimes small bright spots are seen scattered here and there over the disk.

Generally the details on the surface of Jupiter change from day to day, so that no permanent map of the planet can be made. If an exact drawing of the planet is made on one evening, it will be found, two or three evenings later, when the same side is presented to us, that many changes have taken place. Sometimes, however, spots are seen which remain for weeks, months, or even years. Most remarkable

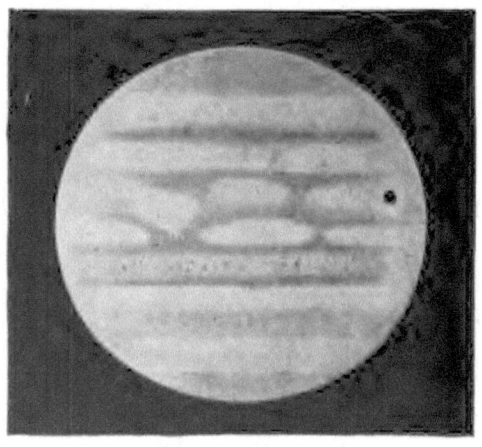

FIG. 88. — Telescopic view of Jupiter with the dark shadow of a satellite crossing over its disk.

of these was a red spot which appeared south of the equator in the year 1878 or 1879. It varied somewhat from time to time, but continued without any material diminution for more than ten years. Then it began to fade out in a varying and uncertain way, being sometimes conspicuous and sometimes seen with difficulty. Since 1892 it has been only occasionally visible.

From the variability of the surface, we conclude that what we see on Jupiter is not land, and probably not water. When carefully examined, the edge of the disk is not sharply defined, but appears much softened, the illumination there being far less than it is at the surface. All this leads us to surmise that what we see on the surface of Jupiter are great banks of clouds floating in an atmosphere, and carried about by strong winds blowing mostly in an east and west direction. These clouds are so thick that we cannot see the body of the planet through them. Hence it is not known whether this body is solid or liquid. It is sometimes supposed to be like the sun, an extremely hot mass, liquid or vaporous, compressed by the weight of its outer portions.

2. The Satellites of Jupiter. — When we look at Jupiter through a telescope, or even through a good spyglass, we shall generally see three or four bright objects near him. These are his satellites, which were discovered by Galileo, and became at once objects of the greatest interest to astronomers. Galileo saw that these objects revolved around Jupiter as the moon revolves around the earth, and the planets around the sun. At that time the doctrine that the earth and planets revolved around the sun was not generally held. But Galileo believed in it, and saw in the revolutions of the satellites an additional strong proof, by analogy, of the revolution of the planets. It is said that one astronomer thought the satellites were illusions produced by the telescope itself. There is also a story of a philosopher who refused to put his eye to the telescope lest he should see the satellites, and be convinced. He died shortly afterward. "I hope," said Galileo, "that he saw them on his way to heaven."

It has sometimes been a question whether these bodies cannot be seen without a telescope. It is certain that if Jupiter were out of the way they would be visible to the naked eye. It seems likely that, when two of them happened to be close together, they have been seen as a single satellite, in spite of the glare of Jupiter.

A fifth satellite, nearer to the planet than the other four, was discovered by Barnard at the Lick Observatory of California in 1892. It is, so far as known, the faintest and most difficult satellite to see in the solar system, being visible only in the most powerful telescopes. Even with such a telescope a well-trained eye is necessary. The four largest satellites are probably about the size of our moon, or a little larger. They are so much fainter than the moon, not only on account of their greater distance from us, but because of their fainter illumination by the sun, being five times farther away from the latter than we are.

Eclipses and Transits of the Satellites. — It is evident that a planet, being an opaque body illuminated by the sun, must cast a shadow, as our earth does. Whenever a satellite enters this shadow it undergoes an eclipse like one of our moon. In the case of Jupiter's satellites these eclipses can be easily observed. The inclination of their orbits to that of the planet is so small, and the planet itself is so large, that all the satellites except the outer one pass through the shadow of the planet at every revolution. Accordingly, if we view one of these bodies when it is going to pass behind the planet, we shall often see that, before it reaches the planet, it fades out and disappears from view. This is because it is entering the shadow. At other times it can be seen coming into view as it emerges from the shadow.

These eclipses are interesting subjects of observation, as they can be seen with quite a small telescope. Their times are predicted in astronomical ephemerides, so that any observer with such a publication can observe these eclipses very frequently if the planet is not too near the sun.

Frequently, also, the satellites pass between us and the planet Jupiter. This is a phenomenon which it is very interesting to watch. When the satellite first enters upon the disk, it commonly appears bright on the darker background of the planet; but, as it approaches the center of the disk, it frequently looks darker than the disk. This is because the center of the disk of Jupiter is brighter than the edge.

THE FOUR OUTER PLANETS

When a satellite is thus passing across the disk, we can frequently see its shadow traveling over the disk near it. This is also a very interesting phenomenon to watch, but for this purpose we need a good-sized telescope. (See fig. 88.)

3. The Planet Saturn. — Among the planets Saturn is next to Jupiter in size, its mass being about one third that of the giant planet. It revolves around the sun in $29\frac{1}{2}$ years. Although smaller than Jupiter, it has about three times the mass of the six smaller planets put together.

Saturn has many points of resemblance to Jupiter. These are: —

1. Its rapid rotation on its axis. The time of rotation is 10 h. 14 m., less than 20 minutes greater than that of Jupiter.

2. Its small density, which is even less than that of water, and, so far as we yet know, less than that of any other planet or satellite.

3. The cloudlike aspect of its surface. But these cloud forms are far less distinct than they are on Jupiter, and so can be recognized only with difficulty.

4. The number of its satellites, it having eight or nine in all, three or four more than Jupiter.

Saturn is redder in color than Jupiter and not so bright. The star which it most resembles in appearance to the naked eye is Arcturus.

4. The Rings of Saturn. — When seen with a large telescope, Saturn is the most wonderful object in the solar system, owing to the magnificent rings which surround it. The appearance of these rings is shown in the picture of the planet. They are perfectly flat, and so thin that, when seen edgeways, they nearly or quite disappear, even in powerful telescopes.

They appear to be two in number, separated by a very narrow dark line, as shown in the figure. The outer ring, it will be noticed, is much narrower than the inner one. Near the two ends, which are called its *ansæ*, a little dark shading is seen on it. It is not yet certain whether this shading indicates a division of this ring into two others, or whether it arises from this part of the ring being composed of darker matter than the remainder.

The inner ring is brightest near its outer edge, and grows darker toward the planet. Its inner portion is so dark as not to be visible except in a pretty large telescope. When this portion was first noticed, it was thought to be a separate ring,

FIG. 89. — Telescopic view of Saturn and its rings.

and was therefore called the crape ring or dusky ring. It is now, however, found to be joined continuously to the rest of the inner ring.

The rings are inclined to the plane of the planet's orbit by about 28°. The direction of their plane remains nearly unchanged as the planet revolves around the sun. Hence, in some positions in the orbit, we see the rings considerably inclined, as shown in the figure. As the planet moves around, the rings appear to grow narrower, through the greater obliquity at which we see them. At length they close down into a mere line, and may quite disappear. As the planet goes on, we see the other side of the rings and they gradually open out

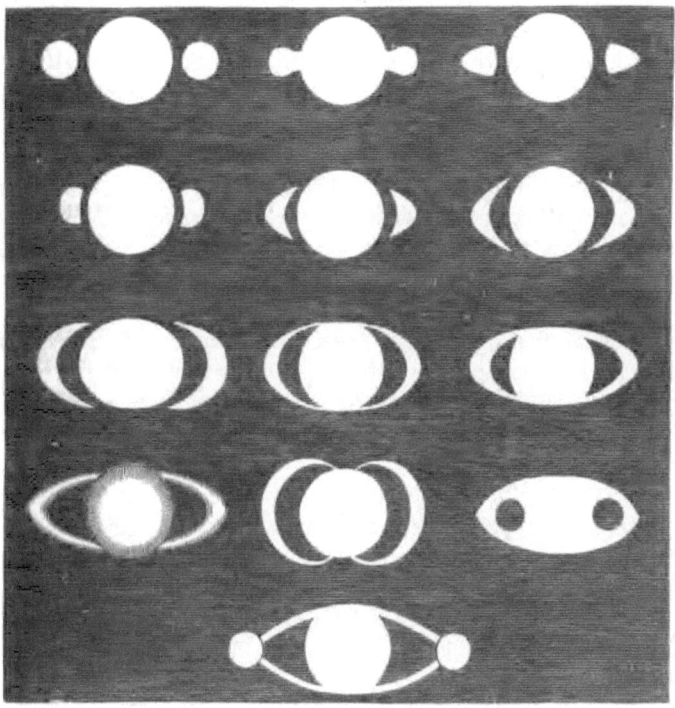

FIG. 90. — Drawings of Saturn and its rings made by various astronomers between 1610 and 1656, before they could see with their poor telescopes what the rings were.

again. The disappearance occurs at every half revolution, or about once in fifteen years.

When these rings were first seen, they were a great puzzle to the astronomical observers, who could not distinguish their true shape. To Galileo they seemed like two little handles to the planet, because he could see only the parts which projected. After two or three years Saturn moved in such a position that the rings were seen edgewise, then they disappeared entirely from Galileo's view. It is said that the philosopher was greatly perplexed by this, not knowing how a celestial body could vary in this way, and feared that something might be wrong in his observations. We give a figure (page 169), showing some of the drawings made by observers during the years between 1610 to 1656. In 1656 the true form of the object was made known by the celebrated Huyghens.

The question of the constitution of the rings was long a difficult one because it was impossible to conceive how immense solid bodies, as they were supposed to be, could be sustained without falling on the planet. But it is now known that they are not solid bodies at all, but only clouds of small particles, or rings of dust or vapor. These little objects are thousands and perhaps millions in number, each revolving in its own orbit. They seem to us to form a continuous surface, owing to their great number.

5. The Satellites of Saturn. — Saturn has eight satellites, (perhaps nine) revolving around it, a larger number than any other planet. The brightest of these was discovered by Huyghens, in 1655. A few years later, Cassini, of Paris, discovered a second one. As the telescope was improved, new ones, nearer to the planet, were added. The nearest one of all, which revolves near the outer edge of the ring, was discovered by Sir William Herschel. The faintest and most difficult of the eight to see, called Hyperion, was discovered by Bond, at Cambridge, in 1848. A ninth satellite, more distant than any of the others, was thought to be photographed by W. H. Pickering at the Harvard Observatory, in 1899, but its periodic time is not ascertained. The following table gives the names of the satellites, their distances from the planet in radii of the planet, their discoverer, and the date of discovery: —

THE FOUR OUTER PLANETS

No.	Name	Distance from Saturn	Periodic Time	Discoverer and Date
1	Mimas	3.3	0 d. 23 h.	Herschel in 1789
2	Enceladus	4.3	1 d. 9 h.	Herschel in 1789
3	Tethys	5.3	1 d. 21 h.	Cassini in 1684
4	Dione	6.8	2 d. 18 h.	Cassini in 1684
5	Rhea	9.5	4 d. 12 h.	Cassini in 1672
6	Titan	20.7	15 d. 23 h.	Huyghens in 1655
7	Hyperion	26.8	21 d. 7 h.	Bond in 1848
8	Japetus	64.4	79 d. 22 h.	Cassini in 1671

There seem to be two gaps in the distances, one between Rhea and Titan, the other between Hyperion and Japetus. We might suppose that there are little satellites in these gaps, but, if so, the most careful search with the most powerful telescopes has failed to reveal them.

These objects are of very unequal brightness. The innermost one, Mimas, and the seventh, Hyperion, can be seen only with large telescopes. Enceladus is nearly as difficult as Mimas. Titan can be seen with a very small telescope. The others are intermediate in difficulty.

All these satellites, except the outer one, revolve in the plane of the ring. Consequently when the edge of the ring is turned toward the earth, the satellites seem to swing from one side of the planet to the other in a straight line, running along the ring like beads on a string. The planes of the orbits are kept together by the attraction of the rings and satellites on each other.

Japetus, the outer satellite, has a remarkable peculiarity. It is much brighter when seen west of the planet than when seen east of it. The explanation of this is that the satellite, like our moon, always presents the same face to the planet, and that one hemisphere is much whiter in color than the other. The result is that the white hemisphere is turned toward us when it is on one side of the planet, and the dark one when it is on the other side.

6. Uranus and its Satellites. — The distance of Uranus from the sun is about twice that of Saturn. When near opposition it shines as a star of the sixth magnitude. It can therefore be seen with the naked eye if one knows exactly where to look for it.

In a small telescope this planet looks like a star. But when a magnifying power of several hundred is used, it will be seen to have a disk. It has a greenish tint, and the spectroscope shows that it is surrounded by a very dense atmosphere. Its time of rotation on its axis is unknown.

Uranus was discovered by Sir William Herschel in 1781. He at first supposed it to be a comet. After a few weeks its motion showed that this could not be the case, and its true nature was then soon learned. Herschel proposed to call it the *Georgium Sidus*, or the Georgian Star, after King George, who had afforded him the means of making his discoveries. This name was not used outside of England. Some astronomers proposed to call the planet Herschel, after its discoverer, but this name did not meet with favor either. The name Uranus was then chosen and permanently adopted.

Satellites of Uranus.— A few years after the discovery of the planet, Herschel found that it was accompanied by two satellites. Afterward, he thought he found four others, making six in all. It was therefore supposed for a long time that Uranus had six satellites. But recent investigations with the more powerful telescopes of our time have shown that the four smaller supposed satellites did not belong to Uranus, but were probably fixed stars which Herschel had from time to time seen in the neighborhood of the planet. But two additional very faint satellites were discovered by Lassell quite near the planet. The existence of these has been well authenticated by recent observations. There is probably no great difference in the actual brightness of the four satellites; but it requires a powerful telescope to see any of them, and those nearer to the planet are harder to see than the two distant ones, because of the glare of the planet.

THE FOUR OUTER PLANETS 173

These satellites have one remarkable peculiarity. The orbits of the satellites of the earth, Mars, Jupiter, and Saturn are but little inclined to the plane of the ecliptic; but those of Uranus are nearly perpendicular to this plane. That is to say, the plane in which they revolve is not slightly inclined, like the other planes we have described, but is tipped up at almost a right angle. Consequently, when the planet is in certain parts of its orbit, we see these orbits almost perpendicularly, and can watch the satellites in their whole course around the planet. As Uranus revolves in its orbit, the plane of the orbits of the satellites retains its direction. Consequently, twice in every revolution of Uranus this plane passes through the position of the earth. The motions of the satellites then seem to take place in a nearly north and south direction, swinging on each side of the planet.

7. Neptune and its Satellite. — Neptune is, so far as we know, the outermost planet of our system. Its discovery was one of the most remarkable events in the history of astronomy, illustrating the great exactness of astronomical theory and the power of the human intellect in penetrating the mysteries of the celestial motions. Its existence was predicted and its direction from the earth made known before it had been recognized by the human eye.

About forty years after the discovery of Uranus, tables for calculating the motions of that planet were made by Bouvard of Paris. Such tables enable us to calculate not only how the planet would move in an elliptic orbit under the influence of the sun's attraction, but also to what extent and in what direction it is drawn away from this elliptic orbit by the attractions of the other planets. Soon after Bouvard's tables were published, it was found that the planet did not move exactly in accordance with them. The deviation was indeed very small, even when at its greatest. How small it was, we may judge by the fact that, if there had been two planets in the heavens, the one in the position of the real Uranus, and the other in the position where Bouvard's tables said the planet ought to be, the two would have seemed to the naked eye as a single star. But in a telescope this small difference was very perceptible, and the deviations were a source of surprise to observers.

About 1843 it occurred to several astronomers that these deviations were probably due to the planet being drawn from its place by the attraction of some unknown planet, probably one whose orbit lay outside that of Uranus. Two mathematicians, Leverrier in Paris and Adams in England, thereupon undertook to calculate where this unknown planet should be situated and how it should move in order that its attraction might produce the observed motions of Uranus. The two men agreed remarkably in their conclusions. The difference in the directions which they assigned to the unknown planet was only one or two degrees.

Leverrier knew that the astronomers of Berlin were engaged in making maps of the stars in that part of the sky where the planet should be. It occurred to him that it might be found by the aid of such a map. He therefore wrote to Doctor Galle in the autumn of 1846, asking him to examine the map and compare it with the heavens. Pointing his telescope in the required direction, Galle found an object which was not on his map of the stars. But he could not tell certainly whether it was a star until the following evening. Then he looked again and found that the object had changed its position. This showed that it was really a planet; and it proved to be the one which had been predicted. This result was of the greatest interest, and most of the astronomers of the world who had instruments at their disposal began to watch the course of the newly discovered body.

Satellite of Neptune. — Mr. Lassell of England, soon after the discovery of Neptune, noticed a very faint star near it. After watching for a few evenings, this star proved to be a satellite. It is the only one that Neptune is known to have. Its orbit is inclined to the ecliptic about 30°. But its motion has the remarkable peculiarity of being retrograde instead of direct. That is to say, if we should look down upon the solar

THE FOUR OUTER PLANETS

system from a great height in a direction perpendicular to the plane of the ecliptic, we should see all the planets and satellites making their revolutions in the opposite direction from that of the hands of a clock, the satellites of Uranus and Neptune excepted. Those of Uranus, as I have just explained, would appear to move nearly in the same plane in which we were looking down on them, swinging back and forth on each side of the planet, but that of Neptune would move in the opposite direction from all the others.

The orbit of this satellite changes its position in such a way as to show that the planet is spheroidal in form. It is therefore concluded that it has a rapid rotation about its axis. But it is not possible to detect this rotation with a telescope, owing to the great distance of the planet.

CHAPTER XIII

COMETS AND METEORS

1. Appearance of a Comet. — Comets are objects of unusual aspect which, to the ordinary observer, sometimes seem to hover in the heavens for a few weeks or months, and then gradually disappear. Occasionally one of these objects is so large and brilliant as to command universal attention. Smaller ones, that would hardly be noticed unless the attention were called to them, are more frequent. Smaller ones yet, that cannot be seen except with a telescope, appear in such numbers that a year seldom passes without several being observed.

Parts of a Comet. — A large comet consists of three parts, the *nucleus*, the *coma*, and the *tail*. These parts are not completely distinct, as one merges gradually into the other.

The *nucleus*, to the naked eye, looks like a star. It would not excite notice but for the coma and tail by which it is accompanied.

The *coma* (Latin for hair) is a mass of cloudy or vaporous appearance in which the nucleus is enveloped and which shades off so gradually that we cannot say precisely where it ends. It gives the nucleus the appearance of a star shining through a little bunch of fog. The nucleus and coma together are called the *head* of the comet.

The *tail* is a continuation of the coma, and consists of a stream of foggy or milky light, growing broader and fainter as it recedes from the head, until the eye can no longer trace it.

The direction of the tail is always away from the sun. Its extent is very different in different comets. In some of the largest it extends over a considerable arc of the heavens, while in others it is comparatively short. Its actual length is nearly always many millions of miles.

Variety of Aspects of a Comet. — Comets that cannot be seen by the naked eye are called *telescopic comets*. These objects frequently have no tail, and occasionally the nucleus is so faint as to be scarcely visible even in the telescope. In these cases the comet looks like a minute patch of fog.

A *periodic comet* is one which is known to perform a regular revolution round the sun, like a planet. These comets move in much more eccentric orbits than planets.

Most of the periodic comets make their revolution in a few years, between three years and fifteen. But there are also a few comets having periods ranging from seventy years upward. Only two or three of these have ever been seen at more than one return.

2. Comets belong to the Solar System. — It is now known that all comets may be considered as belonging to the solar system. They are attracted by the sun as the planets are; but instead of moving in nearly circular orbits, like the planets, they drop down, as it were, from an immense distance, generally from a region far outside the orbit of Neptune. If one fell exactly toward the sun, it would drop into it. But every comet hitherto known has its motion directed a little one side of the sun or the other. As it drops it goes faster and faster until, when it gets very near the sun, it is moving with such rapidity that the sun can no longer hold it. It whirls around and goes off again, nearly in the direction from which it came. The ellipse in which it is supposed to move is so elongated that we cannot see the comet in any part of it except quite near the sun.

Comets are entirely invisible except in that small part of their orbit nearest to the earth and sun. As one is dropping toward the sun, some keen-eyed astronomer, with a little telescope, is almost sure to detect it if the earth is in a favorable position for seeing it. He then announces his discovery by telegraph to his fellow astronomers in Europe and America, telling exactly where it is to be seen. When first found it commonly looks like a mere patch of fog. Then as it gets nearer and brighter the nucleus is seen, and if the comet is a big one, a little tail begins to form. After the comet whirls around the sun it again grows smaller and fainter, the nucleus gets dim, the tail shortens, and when it is lost sight of in the telescope it looks much as it did when it was first seen.

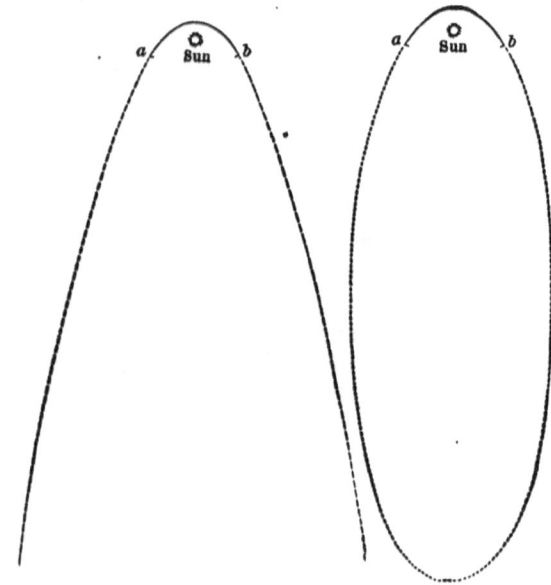

Fig. 91. — Showing two cometary orbits, the one an elongated ellipse, the other extending out we know not how far. The arc *ab* shows the portion of each orbit within which it is possible for us to see the comet. These arcs are so near alike that it is often impossible to distinguish between them.

3. Orbits of Comets. — You may ask how it is that astronomers know anything about the movement of a comet after it disappears from their telescope. The answer is that they calculate its orbit by the theory of gravitation. Its position in the heavens is observed with the greatest exactness from day to day, and thus it may be known how fast it is moving at any particular point of its orbit where it is visible. In this way it may be calculated how far it will go before the attraction of the sun can stop it and bring it back again. There is at each distance from the sun a certain velocity which the comet might acquire if it fell from an infinite distance. If the comet's velocity exceeds this the sun could never stop it, but it would move away into infinite space, forever. This velocity is less the farther the comet is from the sun. At the distance of the earth's orbit the limiting velocity is about twenty-six miles a second. If it should pass the earth's orbit with a higher speed than this, we should know that it would never return.

Commonly the velocity is so near this limit that it is impossible, from the velocity alone, to be quite sure whether it will return or not. But as there is no evidence of a comet ever exceeding this velocity, it is believed that every comet belongs to the solar system and will return some time. In the large majority of cases, however, the return will not take place for several thousand years.

How a Comet may have its Orbit Changed. — Very often a comet, in falling toward the sun, or leaving it again, passes so near some great planet, generally Jupiter, that the attraction of the latter completely changes its orbit, by retarding the comet so that the sun can hold it in a smaller orbit. Then the comet will describe an ellipse around the sun with a period of a few years. It is probable that several of the comets of short period have had their orbits changed in this way by the action of Jupiter.

Twenty or thirty periodic comets are known, and new ones are made periodic from time to time, in the way just described. The last case of the kind was that of a comet which appeared in 1889, known as Brooks's comet, after the astronomer by whom it was first seen. This body is still revolving in a period of about seven years. It returned in 1896 and will return again about 1903, 1910 and so on.

When a comet has thus been made periodic by the attraction of Jupiter, it is liable to meet that planet again at some future time and to have its orbit again changed. Its period may be shortened again, or lengthened, or the planet may give it a swing that will send it off from the sun altogether, so that we shall never again see it. This happened to a comet discovered by Lexell in 1770.

4. Remarkable Comets. — In former times comets excited great fear because they were supposed to portend pestilence, war, the death of kings, or other calamitous events. Naturally the more striking and brilliant the comet the greater the terror thus excited. But when the astronomers found that these bodies moved according to the law of gravitation, and had no power of their own, these fears vanished.

Halley's Comet. — This comet is one of the most remarkable in history; not because it was very bright, but because it was the first one found to be periodic. It was seen in 1682. Halley, an eminent English mathematician, computed the position of the orbit and found that it moved in nearly the same orbit as a comet observed by Kepler in 1607. He therefore suggested that it might come back about every 75 or 76 years. Subtracting 76 years from 1607 brings us back to 1531. In this year a comet had actually been seen. Again subtracting 75 years carries us back to 1456. It is known from history that in that year a comet appeared which caused such terror that the Pope ordered prayers to be offered for protection against the Turks and the comet. It is supposed that this gave rise to the popular myth of the Pope's bull against the comet.

Halley now ventured the prediction that this object would return again about 1758. During this time the mathematical methods of calculating the motion of such a body were invented. Clairaut, one of the first mathematicians of France, calculated the effect which would be produced by the attractions of Jupiter and Saturn, and found that the comet would be so delayed that it would not reach its perihelion until about April, 1759. It actually did pass its perihelion on March 1, 1759, so that Clairaut was within a month of the truth.

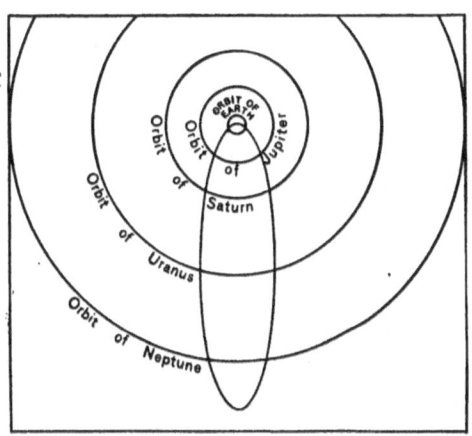

Fig. 92. — Orbit of Halley's comet crossing orbits of planets.

Seventy-six years more were to elapse, when the comet would again be expected. During this time great improvements were made in the methods of computing the attraction of the planets on such a body. So successful were the astronomers in this work that, long before the comet was seen, they predicted that it would pass its perihelion early in November, 1835. It was seen three or four months before this time, and actually did pass the perihelion only three or four days after the time of the best prediction. The comet was followed until May 17, 1836, when it disappeared from the sight of the most powerful telescopes of the time and has not been seen since. But the astronomer can follow it with the eye of science almost as certainly as if he saw it in his telescope.

It passed aphelion, outside the orbit of Neptune, about 1873, and has been on its way back ever since. An exact calculation of its return has not yet been made, but it is expected about the year 1911.

The Great Comet of 1843. — This comet burst suddenly into view in the neighborhood of the sun toward the end of February, 1843. What was most remarkable was that it was visible in full daylight, so that some observers actually measured the angle between it and the sun. It was watched until the middle of April, when it disappeared. It passed nearer the sun than any other body was ever known to pass, — so near that with a very slight change of its original motion, it would actually have struck the sun. So far as observations show, it will not return again for several hundred and perhaps several thousand years.

Fig. 93. — Telescopic view of the head of Donati's comet.

Donati's Comet of 1858. — This comet was one of the most magnificent on record. It was discovered by Donati, of Florence, on June 2, 1858, and, when first seen, seemed to be merely a cloudy patch of light. No tail was noticed until the middle of August, and, at the end of that month, the comet itself was barely visible to the naked eye. As it approached the sun, in September, it brightened up with great rapidity. During the first half of October, its tail was 40° in length, and of a curious featherlike form.

The period of this comet is found to be 1850 years. It will therefore return again about the thirty-ninth century.

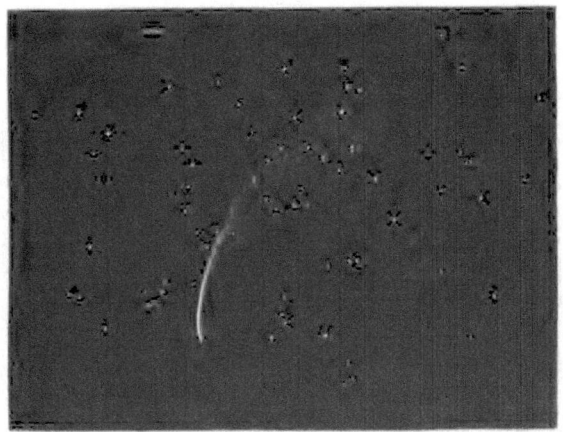

Fig. 94. — Naked-eye view of Donati's comet.

The Great Comet of 1882. — The last very bright comet which appeared was that of the year 1882. It was remarkable, not only for its size and splendor, but from the fact that it moved very nearly in the orbit of the comet of 1843, swinging dangerously near the sun, as that comet did.

Among the comets of short period, and those whose history is most interesting, are those of Encke and Biela.

Encke's Comet. — This comet was first recognized as periodic in 1818, when it was found by the astronomer Pons of Marseilles, France. On computing its orbit he found that it had been seen three times before, the first time in 1786. But the former observers did not know that it was the same comet which had reappeared. It was now found to be revolving round the sun in a period of about 1200 days, or between 3

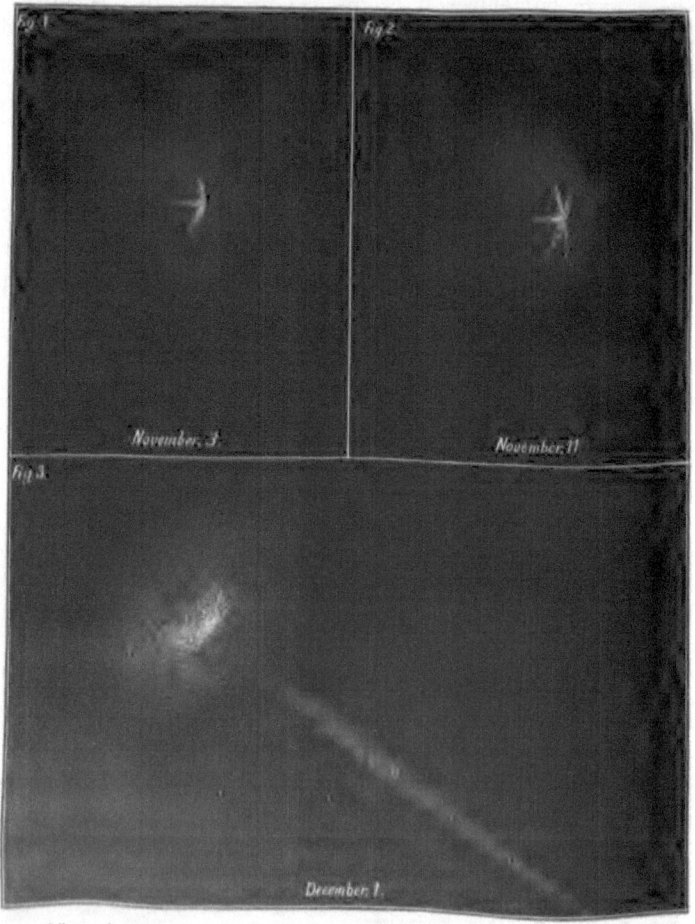

Fig. 95. — A telescopic comet. Three views of Encke's comet as drawn by Vogel in 1871.

and 4 years. It has been followed from time to time, during most of its returns to perihelion, ever since. The most thorough investigation of its orbit was made by the German astronomer Encke, whose name, for this reason, is given to the comet.

The most remarkable discovery of Encke was that the comet seems to be gradually shortening its period of revolution. At most of its returns it seems to be accelerated a few hours. This is supposed to arise from the comet passing through some extremely rare medium which resists its motion when it passes nearest to the sun. In consequence its orbit is growing smaller, and thus the revolution is completed in a shorter time. As no other comet shows similar acceleration, it is not regarded as quite certain that a resisting medium is the real cause of the change.

Biela's Comet. — The early history of this comet is a good deal like that of Encke's. It was seen in 1826 by an Austrian named Biela. On computing its orbit it was found to be periodic. On calculating its motions back it was found to have been observed in 1772, and again in 1805. But both the previous observers thought the comet to be new.

Its time of revolution was now fixed at 6 years and 8 months. In 1846 it again returned to perihelion, and was found to have suffered an accident never before known in the case of a heavenly body. It had separated into two parts, so that, instead of one comet, two were seen.

At its next return, in 1852, both parts were observed. But the comet has never been seen since, though repeatedly looked for at the proper times. Its parts have all been dispersed, but are doubtless revolving round the sun as minute invisible particles, as we shall presently explain.

5. Constitution of Comets. — From what we have said, it is plain enough that the coma and tail of a comet are neither solid nor liquid, but are composed of an exceedingly thin vapor, lighter and thinner than the finest vapor on the surface of the earth. This vapor seems to rise from the nucleus of the comet under the influence of the sun's rays, much as heat makes a pot of water boil. But the vapor is enormously thinner than any steam rising from a pot. It forms a cloud around the nucleus, and moves along with it. As the comet approaches the sun, the vapor is, in some way not yet fully understood, repelled by the sun, and thus driven off in a stream. This stream is the tail of the comet, and is always directed away from the sun, thus showing that the sun repels instead of attracting the matter that composes it. Hence the tail is not an appendage which the comet carries with it, like a bird carries its tail, but is rather like a stream of smoke rising from a chimney.

It follows from this that a comet which shows a real tail is, when near the sun, continually losing its substance by evaporation. When we consider how very large the tail is, nearly always many millions of miles in length, we might suppose that the matter of the comet must be evaporating very fast. But the actual amount of matter in the tail of a comet is exceedingly small. The tail is so tenuous that the stars can be seen through millions of miles of it without any diminution of their light. It reflects so little of the sun's light that it cannot be seen at all in the daytime or in bright twilight. It is quite possible that there may be only a few pounds of matter in the whole tail of a comet.

Still, all the matter that can evaporate from the comet will in time be lost. Thus it happens that the tails of the periodic comets gradually diminish as they make their successive revolutions. This is because the volatile matter is gradually being evaporated. Those comets which have made a great many revolutions scarcely show any tail at all.

One result of this is that periodic comets, like men, have only a limited time in which to exist. They grow thinner and paler as they make their revolutions, and are probably nearly all doomed at some time to disappear. Indeed, several have disappeared during the last 50 years. But new ones are being brought in by the attraction of Jupiter to take their places in the way already described, so that the number is as great as ever.

6. Meteors. — Every one who looks at the heavens with care by night must notice the meteors, or shooting stars, of which several can be seen on any clear night of the year. These objects suddenly come into view, like a moving star, dart swiftly along for a moment, and then disappear. Their cause has only recently been discovered.

It is now known that, besides the bodies of the solar system which we see with our telescopes, there are countless millions of little particles of matter revolving around the sun in orbits of various forms and sizes. These particles are so small that no telescope is powerful enough to show them. As the earth flies along in its orbit it is constantly encountering these little particles, which, of course, first strike the atmosphere by which the earth is surrounded. Now it is a law of physics that, when a body moves with exceeding swiftness through the air, heat is generated by the friction and resistance. The great velocity of these particles moving around the sun — a velocity of many miles a second — results in so much heat and friction by the air that the particle is almost instantly destroyed or burnt with a great evolution of light and heat. Then those who see this light call it a meteor or shooting star. The invisible particles, as they exist before striking the air, are called *meteoroids*.

The meteoroids are very different in size. Hence shooting stars are of very different degrees of brilliancy. Sometimes the meteoroids are so large that they rush a long distance through the air before they are burnt up by the fervent heat. Then we see a very brilliant meteor, and sometimes hear a loud report like thunder, or the discharge of artillery. This is caused by the concussion of the air by the meteoroid.

Occasionally the latter is so large that it falls to the ground before being burnt up. It is then called a *meteorite*. If examined immediately after it falls, it is found to be very hot on the surface, perhaps partly melted, in consequence of its rapid passage through the air. Many meteorites have fallen in this way and some of them are preserved in our museums. They are found to consist very largely of iron and stone.

It is not certain whether the meteoroids which form the ordinary shooting stars are composed of the same substance as these large masses that fall to the ground. When they disappear in the high regions of the air, they are completely dissipated, so that it is impossible to find any remains of them.

7. Meteoric Showers. — Sometimes shooting stars appear much greater in numbers than usual. They may follow each other so rapidly that they can scarcely be counted. Then there is said to be a *meteoric shower*. Several remarkable showers of this kind are recorded in history. One occurred in the year 1799 and another in the year 1833. A third, less striking, was seen in 1866. All occurred at the same time of the year, about November 14.

It is now known that these showers are caused by an immense stream of meteoroids which revolve around the sun in a very elongated orbit, making about three revolutions in a century. The swarm is so long that although millions are passing all the time, it takes two or three years for the whole procession to pass a given point. It happens that the orbit of this swarm intersects the orbit of the earth at that point where the earth passes about the middle of November of every year. Thus, during the two or three years that the swarm is passing, the earth will encounter it two or three times in succession. Then, when the swarm has got by, nothing more will be seen of these meteors, except perhaps a few stragglers.

It is known that a comet is revolving in the same orbit with these meteoroids that cause the November showers. Hence it is supposed that the meteoroids belong to the comet and were once a part of it.

Radiant Point of a Meteoric Shower. — Each meteor in a shower of course describes a certain path on the celestial sphere. If we continue each of these paths in the reverse direction to that of the motion of the meteor, we shall find them all directed from one and the same point in the heavens. This is called the *radiant point* of the shower.

FIG. 96.—The radiant point of a meteoric shower.

The radiant point is an effect of perspective. The real paths described by the meteors are parallel and nearly straight lines, which, however, look to us like circles on the celestial sphere. The radiant point is simply that point of the sphere from which the lines are directed.

The August Showers. — Although the November showers which we have described are the most remarkable recorded, it is found that there are other nights of the year in which meteors are seen in unusual numbers. One of these dates is about the 9th or 10th of August. About this time every year, if one sits up until midnight, he will see an unusual number of shooting stars, which mostly move from the northeast toward the southwest. They are distinguished by leaving trails behind them which last for a few seconds. They are called *Perseids* because their radiant point is in the constellation *Perseus*.

CHAPTER XIV

THE CONSTELLATIONS

1. About the Stars in General. — In the most ancient times, the priests and astronomers who watched the sky by day and night saw that the heavenly bodies were of two kinds in respect to their motions. Seven of them, the Sun, the Moon, Mercury, Venus, Mars, Jupiter, and Saturn, seemed to have separate motions of their own. Hence they were called planets, from the Greek word *planetes*, a wanderer. This is the origin of the word planet. The remainder, of which one or two thousand were easily seen, all seemed fixed in the celestial sphere. This sphere was supposed to turn on its axis every day, carrying these bodies with it. This is why the latter were called fixed stars. The fixed stars comprise all the stars we see in the heavens at night, the planets excepted.

We have already shown that the seven planets of the ancients belong to the solar system, while the stars are many thousand times farther from us than the planets are. Let us try to gain an idea of these distances. There are in the heavens three or four pairs of stars so close together that only a keen eye can see that there are two stars. If a railway train could run at a speed of 60 miles an hour from one of these stars to the other, without ever stopping, it might require a million of years for the journey. During the 5000 years since the beginning of written history, the journey would hardly have been commenced. It would be as if a train which was to run to a city several miles away was just pulling out of the station.

Each of the stars which has been carefully observed is moving forward in what seems to us a straight line, often at the rate of many miles per second, and the same is probably true of all. This motion of the stars is called *proper motion*. Owing to the great distance of the stars, it cannot be seen except by telescopic observation. It will be described subsequently.

The stars are bodies like our sun, surrounded by layers of heated vapor, and shining by their own light.

They are very different in real brightness. Some are hundreds or thousands of times brighter than our sun; as I have already said, they look small because they are so far away.

They are at very different distances. The most distant that we know are hundreds or even thousands of times farther than the nearest.

Number of Stars. — In the whole heaven there are about 5000 stars that can be seen by an ordinarily good eye. But as one half of these are, at any one moment, below the horizon, and many of the other half so near the horizon that the fainter ones cannot be seen, it is not to be expected that more than 1500 to 2000 can ever be seen at the same time.

For every star visible to the naked eye, there are thousands that can be seen with a telescope, though invisible to the naked eye. These are called *telescopic stars*. With every improvement in the telescope more are seen, so that the total number made known by the largest instruments probably amounts to 50 millions. The number that may be photographed is yet greater, perhaps even 100 millions.

The Milky Way. — Every one who carefully looks at the sky on a clear night must notice the Galaxy, or *Milky Way*, as it is commonly called; that beautiful white arch which spans the heavens. To the naked eye it appears as an irregular row of cloudlike masses, having a milky aspect, from which its familiar name is given it. When examined with a telescope, the light is found to come from innumerable stars, too faint to be seen separately with the naked eye. With a good-sized telescope of a low magnifying power, the field of view is seen to be studded with such stars shining like little diamonds. The total number of telescopic stars which form the Milky Way is

THE CONSTELLATIONS

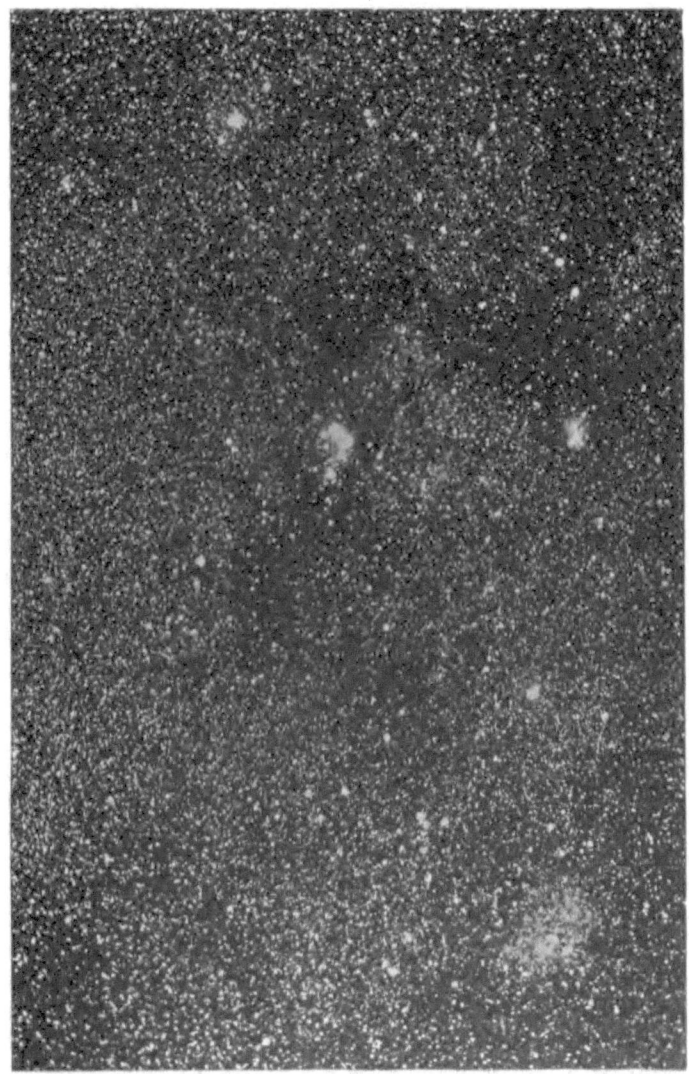

Fig. 96.—The Milky Way near the star 15 Monocerotis. $AR = 6$ h. 35 m. Decl. $= $ N. 10°. (Photographed by Barnard, 1894; exposure, $3\frac{1}{2}$ hours.)

very great, probably greater than the number of all the stars in the remainder of the heavens.

The question how the stars are arranged in the Milky Way is one of the most interesting with which astronomers concern themselves. With the naked eye we can see that they are not scattered uniformly, because if they were the Milky Way would be everywhere equally bright. Many of them are collected into masses or clusters. Some of the clusters contain so many stars that it is almost impossible to count them, because they cannot be separated with the most powerful telescope; in fact, it is hardly possible to sweep a great telescope along the course of the Milky Way without finding collections of beautiful objects too numerous to be separately described.

Magnitudes of the Stars. — The stars are very different in apparent brightness. This arises both from their different real brightness and their different distances.

The ancient astronomers divided all the stars they were able to see into six classes, according to their apparent brightness or magnitude. At one extreme were the fifteen brightest stars. These were called of the first magnitude. At the other extreme were the great number of stars which could barely be seen with a good eye. These were called of the sixth magnitude. Between these extremes came, in regular order of brightness, stars of the second, third, fourth, and fifth magnitudes. The brightest six stars of Ursa Major and the brightest four of Cassiopeia are of the second magnitude. The stars a degree fainter than these are of the third magnitude; those yet a degree fainter of the fourth. The fifth are the faintest that one will easily see, unless the sky is clear and moonless, and his eyes are good.

The same system of magnitudes is continued to the telescopic stars. Thus we have stars of the seventh, eighth, ninth, and so on up to the fifteenth or sixteenth magnitudes. The latter are about the faintest that can be seen or photographed with the most powerful telescope. But we do not know how many still fainter ones may really exist. With every increase

THE CONSTELLATIONS

in the power of the telescope new stars are brought into view.

Colors of the Stars. — If we carefully compare the stars with each other, we shall see that they are perceptibly different in color, ranging from the reddish tint of Aldebaran to the bluish white of Vega. This difference shows that the substances of which these bodies are composed, and also their temperatures, are different. No doubt the atmospheres by which they are surrounded absorb a part of the light from the star, and thus change the apparent color. The red stars are supposed to be less hot than the blue ones, and to be surrounded by a dense atmosphere, which absorbs the blue light, but lets the red light pass.

2. How the Constellations and Stars are Named. — A *constellation* is a large group of stars. A *cluster* is a small group of many stars very close to each other.

The stars are commonly divided among 85 constellations. The majority of these were imagined by the ancient astronomers. The remainder have been mapped out in modern times.

To most of the older constellations are given the names of heroes, goddesses, or animals. The outlines of these persons or objects were supposed to be drawn on the sky in such a manner that the figure should include the principal stars of the constellation. Thus we have *Cassiopeia*, the Lady in her Chair; *Andromeda*, the Chained Goddess; *Auriga*, the Charioteer; *Ursa Major*, the Great Bear; etc.

Special names, mostly given by the Arabian astronomers, are often used to designate the brighter stars. But it is now common to name a star as we do a person, by a Christian name and a surname. The Christian names for the principal stars are the letters of the Greek alphabet, Alpha (α), Beta (β), Gamma (γ), etc. The surnames are the names of the constellations to which the star belongs. They are generally expressed in Latin. This method of naming the principal stars is called the *Bayer system*, after the astronomer Bayer who introduced it about 1600.

Generally, but not always, the brightest star in a constellation has the name Alpha, the next brightest the name Beta, and so on.

Only the principal stars can be named in this way. To others numbers are assigned, according to various systems. The following are the names of some of the principal stars on the two systems. First is given the old name, generally from the Arabic, and then the name on the Bayer system. which is now generally used by astronomers.

Old Name	Bayer Name
Algenib	γ Pegasi
Polaris / The Pole Star	α Ursæ Minoris
Archernar	α Eridani
Algol	β Persel
Alcyone	η Tauri
Aldebaran	α Tauri
Canopus	α Argus
Sirius	α Canis Majoris
Castor	α Geminorum
Procyon	α Canis Minoris
Pollux	β Geminorum
Regulus	α Leonis
Mizar	ζ Ursæ Majoris
Spica	α Virginis
Arcturus	α Boötis
Antares	α Scorpii
Vega	α Lyræ
Altair	α Aquilæ
Fomalhaut	α Piscis Australis
Markab	α Pegasi

When translated these Bayer names would read *Gamma of Pegasus, Alpha of the Little Bear*, and so on, Pegasus, the Flying Horse, Ursa Minor, the Little Bear, etc., being the names of the constellations.

THE CONSTELLATIONS

3. Description of the Principal Constellations. — We shall now describe the most noteworthy constellations, clusters, and other remarkable objects in the heavens which can be seen without a telescope. He who enters on this study should look at the heavens for himself on as many clear and moonless nights as possible. It is well to begin with the constellations around the north pole, because in our latitudes they can be seen on almost every clear night of the year. These are called *Circumpolar* constellations.

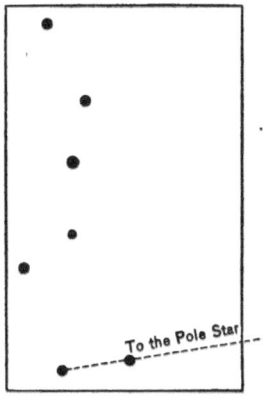

Fig. 97. — The Great Bear or Dipper.

We have already described the two principal circumpolar constellations, *Ursa Major*, or the Great Bear, and *Cassiopeia*. Between these two and in the immediate neighborhood of the pole lies *Ursa Minor*, to which the pole star belongs. It contains two stars of the second magnitude. *Alpha Ursæ Minoris*, or the pole star, is near the end of the animal's tail; *Beta* is in his body. *Draco*, the dragon, is represented as a long, snakelike figure, whose body curves round between Ursa Major and Ursa Minor.

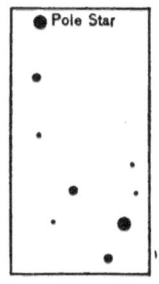

Fig. 98. — Ursa Minor or Little Bear.

The visibility of other constellations will depend upon the time of year and time of day at which we look. We may divide our views into spring, summer, autumn, and winter views. This does not mean that we can get the views only at these respective seasons, but that at these seasons the stars will be in a certain position early in the evening, when it is convenient to look at them. You can see the stars which we call the autumn ones in the spring, if you will get up before daylight in the morning.

4. Constellations visible in the Evenings of February and March.

— In the spring we shall see the Milky Way spanning the heavens, and passing west of the zenith and over to the south. We shall first notice the bright constellations which lie in its course. In the northwest we shall see *Cassiopeia*. Above it is *Perseus*, which can be recognized by a row of stars running along the center of the Milky Way, one of which, of the second magnitude, is much brighter than the rest. This is called *Alpha Persei*.

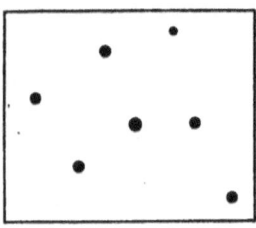

Fig. 99. — Cassiopeia.

In this constellation you will see a remarkable cluster which looks, to the naked eye, like a cloudy mass. This is called the cluster of Perseus. It forms the hilt of the sword which the hero Perseus wears when he is painted on the sky. Even with a good spyglass one can see that this cloudy mass consists of a collection of very small stars.

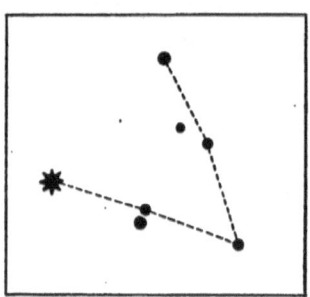

Fig. 100. — Aldebaran and the Hyades.

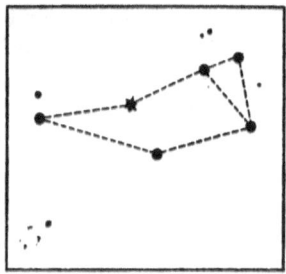

Fig. 101. — The Pleiades, or the Seven Stars.

Above and south of Perseus lies *Auriga*, the Charioteer. It contains a star of the first magnitude called *Capella*, the Goat, which will now be west of the zenith.

West of the zenith and below the Milky Way is the constellation *Taurus*, the Bull, which may be recognized by the bright

THE CONSTELLATIONS 199

red star *Aldebaran,* which is near one eye of the Bull, and used to be called the Bull's Eye. It is at one end of two lines of

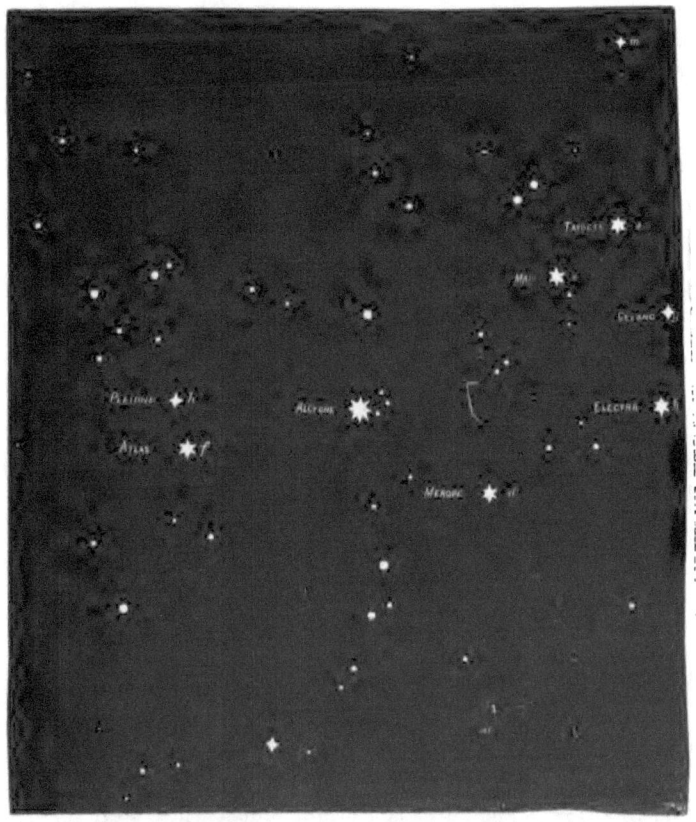

FIG. 102. — Telescopic view of the Pleiades, after Engelmann. The six brightest stars are easily seen by the naked eye; the four or five next in size are also visible to very good eyes.

stars, forming a figure like the letter *V*, called the *Hyades.* One arm of the *V* is a very pretty close row of stars.

In the same constellation, a little to the northwest, you see the *Pleiades,* or "seven stars," a cluster which is familiar to

200   ASTRONOMY

all. In reality only six stars of this cluster can be plainly seen with the naked eye. A good eye on a clear moonless night may see five more, making eleven. There is an old tradition that the cluster originally had seven stars, of which one disappeared. This, however, is probably a myth. With a telescope many additional stars may be seen in this cluster.

East from Taurus, on the other side of the Milky Way and near the zenith, lie *Gemini*, the *Twins*. They are recognizable by two stars of the first magnitude, *Castor* and *Pollux*, each of which marks a head on one of the Twins.

Below Castor and Pollux we see *Canis Minor*, the *Little Dog*, marked by a star of the first magnitude, *Procyon*.

Across the Milky Way from Canis Minor, and west of the meridian, we see *Orion*, the most brilliant constellation in the heavens. It contains two stars of the first magnitude, called *Alpha* and *Beta* on our modern system, but *Betelguese* and *Rigel* on the old system. The former is a reddish star marking the shoulder of Orion. The latter is a bright bluish white star, farther below, marking his left foot. A row of three bright stars between these two marks his belt, and three more in a vertical line below show his sword. His head is formed by a cluster of three little stars to the right of Betelguese.

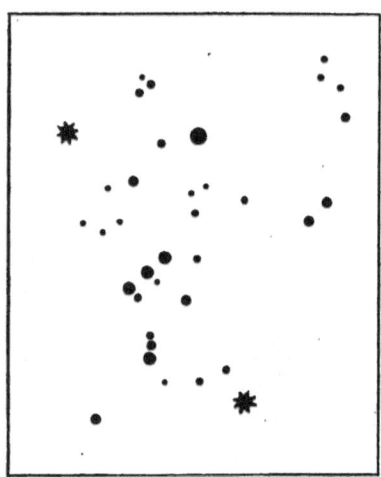

FIG. 103. — Orion.

Southeast of Orion, on the western border of the Milky Way, is *Canis Major*, the Great Dog, remarkable for containing *Sirius*, the brightest fixed star in the heavens.

THE CONSTELLATIONS 201

On the southern horizon will lie *Argo Navis*, the *Ship Argo*, a very large and celebrated constellation, of which the greater part never rises in our northern latitudes. It contains the bright star *Canopus*, the second brightest in the heavens, which in our country can be seen only in the southern states.

Another constellation is *Aries*, the Ram, which will be setting in the west. It may be recognized by three stars of the second, third, and fourth magnitude, forming an obtuse triangle.

East of Gemini will be seen *Cancer*, the Crab, which contains no bright stars, but is remarkable for *Præsepe*, which looks to the naked eye like a little faint spot of milk in the sky. A very small telescope shows it to be a cluster of stars.

Still farther east is seen *Leo*, the Lion, which is marked by the bright star *Regulus*. North of it will be seen a curved row of stars forming a figure like a sickle, of which Regulus is the handle.

5. The Early Summer Constellations. — The next view of the heavens which we shall describe is that in the evening of May, or an early evening of June. The Milky Way passes so near the horizon that it will probably be invisible.

In the west can be seen Castor and Pollux and Procyon, but the other brilliant constellations are near the horizon or below it.

We see Leo, just described, a little west of the meridian.

A little east of the meridian will be seen *Virgo*, the Virgin, in the south, which has a bright star *Spica*, about as bright as Regulus.

Scorpius, the Scorpion, will be low down in the southeast, having just risen. We shall describe it later.

Quite near the zenith we shall see *Coma Berenices*, the Hair of Berenice. This is a very large thin cluster of faint stars, quite irregular in form.

East of this cluster we see *Boötes*, the Herdsman. It is marked by *Arcturus*, a very brilliant reddish star, mentioned in the Book of Job. In the mythological arrangement of the constellations, Boötes is represented as holding two dogs in a leash. They are called *Canes Venatici*, and are chasing Ursa Major round the pole.

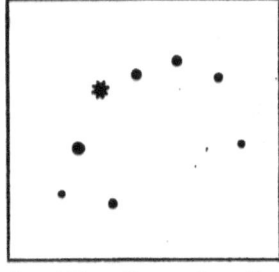

Fig. 104. — Corona Borealis, the Northern Crown.

In the northeast, below Arcturus, we see *Corona Borealis*, the Northern Crown, a small but exceedingly beautiful chaplet of stars, which we can easily imagine to form a crown. The brightest of them is of the second magnitude, and is called Alpha Coronæ Borealis.

6. The August Constellations. — If we look in the latter part of the evening, during the month of August, or in the early evening during the first part of September, we shall again see the Milky Way spanning the heavens like an arch. But we now see a different part from what we saw in the early spring. Arcturus is now sinking in the west, while Cassiopeia is rising in the northeast. In the Milky Way, some distance above Cassiopeia, we see *Cygnus*, the Swan. It has five stars arranged in the shape of a cross, which form the wings, head, and tail of the swan. The brightest of these, Alpha Cygni, is nearly of the first magnitude.

South from Cygnus and near the zenith, is *Lyra*, the Harp, a small but very beautiful constellation. It contains *Vega*, a brilliant, bluish-white star of the first magnitude. South of Vega are four stars of the fourth magnitude forming an oblique parallelogram, by which the constellation can be recognized.

East of Vega, and very close to it, is the star Epsilon Lyræ, which, if you have a very keen eye, you will see to be composed of two stars very close together.

South of Lyra and in the Milky Way, we next look for *Aquila*, the Eagle. It is readily found by the bright star *Altair*, or Alpha Aquila, of the first magnitude, situated between two smaller stars.

Northeast of Aquila is *Delphinus*, the Dolphin, familiarly known as Job's Coffin. Its four principal stars are arranged in the form of a lozenge.

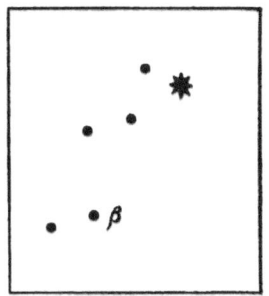

Fig. 105. — Lyra, the Harp.

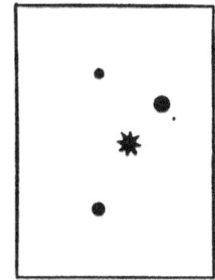

Fig. 106. — Aquila, the Eagle.

Scorpius, the Scorpion, is now low down in the south, west of the meridian. Its brightest star is *Antares*, or Alpha Scorpii, which is red in color and nearly of the first magni-

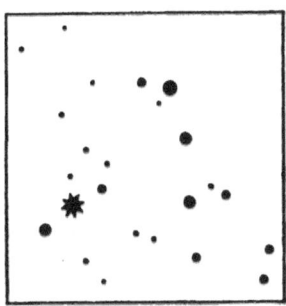

Fig. 107. — Scorpius, the Scorpion, with Antares, the brightest star.

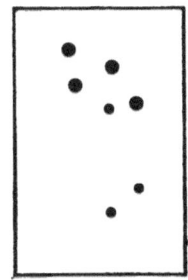

Fig. 108. — Delphinus, the Dolphin, or Job's Coffin.

tude. West of it is a long curved row of stars forming the head and claws of the scorpion.

East of Scorpius we see *Sagittarius*, the Archer.

7. The November Constellations. — We now see the Milky Way a little north of the zenith, seeming to rest on the east and west horizon. All the constellations that we see in its course have already been described.

No bright stars are visible except some of those already mentioned. *Lyra* is in the northwest; *Aquila*, in the west. South of the zenith we see the *Square of Pegasus*, four stars of the second magnitude forming the corners of a large square. Three of them belong to the constellation *Pegasus*, the Flying Horse.

In the south, or southwest, near the horizon, we see the star *Fomalhaut*, belonging to the constellation *Piscis Australis*, the Southern Fish.

The zodiacal constellations, Capricornus, the Goat, Aquarius, the Water Bearer, and Pisces, the Fishes, are all visible in the south or southwest, but they do not contain any very bright stars. Aries, the Ram, is high up, southeast of the zenith, and Taurus, with the Pleiades, is seen lower down in the east; Castor, Pollux, and Procyon, lower yet.

If we wait till ten o'clock, we shall see Orion rise to the south of east, and after him Sirius.

CHAPTER XV

THE STARS AND NEBULÆ

1. The Stars are Suns. — The difference between the apparent brightness of the sun and the stars is such that, in former times, men never suspected that there could be any similarity between them. But the reader who has carefully studied what we have said about these objects will readily understand that the sun is simply one of the stars. In other words, the stars are suns like that which gives us the light of day. They are found to be like the sun in every feature that we can discover.

The first question one would ask on this subject is, How does the brightness of our sun compare with that of the stars? To answer this question we must call to mind on what the brightness of a star depends. We must distinguish between the real brightness and the apparent brightness of such an object. A star of a given real brightness looks fainter to us the farther it is away, just as a distant gaslight looks fainter than one near us. It was formerly supposed that the difference of apparent brightness was due mostly to this cause, the brighter stars being those near us, and the fainter those at a greater distance. But it is found that this is not always the case, and that the stars differ enormously in real brightness. Sirius, the brightest star in the heavens, is many times brighter than our sun. Yet brighter is Canopus, of which we have already spoken. The parallax of this star is found to be immeasurably small, in other words it is immeasurably far away. To shine as brightly as it does it must be thousands of times as bright as the sun.

By comparing the light of the sun and stars, and calling to our aid what knowledge we possess of the distance of the latter, we find that the sun is quite a small body compared with most of the brighter stars in the heavens. The latter are not only suns, but they are suns many times brighter than ours.

In a few cases the mass of a star has been determined. Thus it is found that the mass of Sirius is several times that of the sun. But in this case it is known that the light of the star is greater than that of the sun in a yet greater proportion. It is thus found that many of the stars are much less dense than the sun, being probably like bubbles, masses of very hot gas inclosed in a more or less liquid or cloudy envelope.

This subject is one of which astronomers are now seeking to add to their knowledge. Such knowledge is, as yet, quite certain in only a few cases. Even when certain the result cannot be stated with great exactness. We may compare it to the knowledge of a country which one would acquire by looking around for a few minutes from the top of a high mountain. He would see that some villages were much farther than others, and would know that here was a river and there a hill. But if he tried to make a map of the country he would often err widely in the positions which he assigned to the various objects in the landscape.

2. Proper Motions of the Stars. — We have already said that the stars are generally in motion, often with a very high velocity. But their distance is so great that this motion would not be perceptible to us in a thousand years had it not been determined with astronomical instruments of great precision, especially the Meridian Circle.

Such motions of the stars are called their *proper motions*. So far as yet known these motions take place in straight lines with a speed that never varies.

The proper motion of a star is detected by determining very exactly its right ascension and declination from time to time, and thus finding whether its position changes on the celestial sphere. It is found that nearly all the brighter stars, as well as many of the faint ones, have a proper motion. We may conclude from this that every star in the heavens is really moving. In fact, if a star were really at rest at any moment, the attraction of the other stars would gradually start it moving. In the case of the vast majority of the millions of stars which we see in the heavens, the motion is so slow that it has not yet been detected.

We may express the proper motion of a star in two ways: either as so many miles per second, which would be its real motion; or as such an angle per year or per century, as seen by us, which would be its apparent motion. The real motions are what we should call very rapid, the average being, probably, about 10 miles a second. In a year there are 31,558,149 seconds. Hence, a star moving at this average rate travels more than 315 millions of miles per year. Probably one half the stars are moving straight ahead with this speed, which, so far as we know, never changes, year after year, or century after century. They are traveling forever on a journey of which we do not know either the beginning or the end. Yet, so vast is their distance that only the most refined observations can show how far they move in a whole year. The inconceivable distance we have mentioned is, to our eyes, a mere point in the sky.

3. Motion of the Sun. — The sun, being one of the stars, may be supposed to have a proper motion of its own, as the stars have. In this case it would carry the earth and all the planets with it, without their relative positions being changed. It is just as if a person were walking around a chair in a railway car in motion. He could walk around just as well whether the car were in motion or had stopped.

If a star were at rest and the sun in motion, the star would seem to us to have a motion, on account of the motion of the sun. If we were moving toward a star, then the star would be shown by the spectroscope as if it were moving toward us.

If the sun carrying the earth with it were moving toward the north the star would seem to be moving toward the south. It is, therefore, impossible from any observation on one star, to determine how much of the apparent motion of the star is due to motion of the sun, and how much to actual motion of the star. But if we find that a great majority of the stars seem to be moving in some one direction, we should conclude that this motion is only apparent, being due to a motion of the sun and earth in the opposite direction.

This is found to be actually the case. A large majority of the stars are found to be in apparent motion from the direction of the constellation *Lyra* toward a point in the eastern part of the constellation Argo. This latter constellation, as we have said, is so far south of the equator that it only partly rises above our horizon. We conclude from this that our solar system is moving toward the constellation Lyra. The velocity has been determined in various ways, and is found to be about 10 miles a second.

It is in consequence of this motion of our solar system that so many of the stars seem to be moving away from the constellation Lyra. This apparent motion of the stars is called their *parallactic motion*, because, like parallax, it is due to the change of the direction from which we see them.

The motion of our solar system toward the constellation Lyra is one of the most wonderful conclusions of modern astronomy. One can get an idea of the immensity of the heavens by looking up at the beautiful blue star Alpha Lyra on summer evenings and reflecting that not only during all our lives, but during the lives of all our ancestors for untold generations back, we have been traveling toward it at the rate of about 300 millions of miles per year. And yet the star looks to us as it did to them. Rapid as the journey is, it will probably take our system half a million of years to arrive where the star now is. In the meantime the latter will have moved away, so that it will perhaps be as far away from us as it is now.

THE STARS AND NEBULÆ 209

4. Motions in the Line of Sight. — In recent years the spectroscope has made additions to our knowledge of the motions of the stars which would have been thought impossible before it was invented. Astronomers are now able, by examining the spectrum of a star, to determine how fast it is approaching us, or receding from us. The explanation of the method belongs to the subject of physics, but the general principle is very simple. If the star is approaching us, the spectral lines will all be moved a little toward the blue end of the spectrum; if moving away from us, they will be displaced toward the red end of the spectrum. So, what is done with the spectroscope is to compare the spectrum of a star with that of a substance that gives the same lines as the star. Thus it is seen whether the lines of a star are displaced in one direction or the other, and how far, and thus it is determined whether the star is moving toward or from us, and how fast. The motion toward or from our system is said to be *in the line of sight*.

This is now done by photographing the spectra both of the star and of the substance with which its lines are compared. The measures are then made on the photographic negative.

5. Distances of the Stars. — We have already defined *parallax* as difference of direction, and especially as the difference between the direction of a heavenly body from the center of the earth and from a point on its surface. The distance of the stars is so great that this difference of direction would be

Fig. 109. — Annual parallax of a star.

entirely imperceptible with any instrument that we can make. But as the earth swings round the sun in its vast orbit, 186 millions of miles across, there must be a corresponding change in the direction of the stars from us. This change is called *annual parallax*, because it goes through its course in a year.

When we speak of the *parallax* of a star, we mean the difference in its direction as seen from the sun and from one extremity of the earth's orbit. This is the angle which the radius of the earth's orbit would subtend if seen from the star.

So small is the annual parallax, even of the nearest star, that astronomers were not able to invent instruments which would show it until about the year 1830. Then it was found by Bessel that a small star of the constellation Cygnus, called *61 Cygni*, had a parallax of about $\frac{1}{3}$ of a second. About the same time it was found that the star *Alpha Centauri*, in the southern hemisphere, had a still larger parallax, of which the amount is now known to be about 0.75″.

Since then the parallaxes of about 100 stars have been measured. But in many cases it is so small that there is doubt about the result. The parallaxes of two remarkable ones are supposed to be:

61 Cygni	0.35″
Arcturus	0.03″

Let us now show how, from the parallax of a star, we can determine its distance. Let the little circle *EF* be the orbit of the earth around the sun; *S* the sun, *R* a star. From *R* draw two lines, one to the sun and the other to one extremity of the orbit. The angle *ERS* between these lines will be the annual parallax of the star.

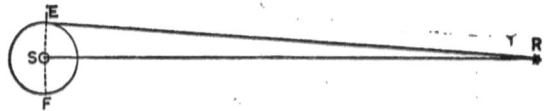

FIG. 110. — Relation between the parallax and distance of a star.

If we draw a circle, as in figure 2, an arc of one degree will be about $\frac{1}{57.3}$ of the radius. That is, an arc of 57.3° will make a length equal to the radius. More exactly, the number of

THE STARS AND NEBULÆ

degrees in the radius is 57.29578. There being 60 minutes in a degree, and 60 seconds in a minute, we shall find by multiplication that there are 206265 seconds in the radius.

Now, if we fancy ourselves to draw a circle with the star as a center and a circumference passing through the sun, as shown in the dotted arc in figure 110 we find, as already stated, that the arc ES or SF, which measures the parallax, is less than one second. Therefore, the distance of a star, or the radius SR, is more than 206265 times that of the earth from the sun. We may find the exact distance by dividing 206265 by the star's parallax. Thus is found: —

Distance of Alpha Centauri	about	275,000
Distance of 61 Cygni	nearly	600,000
Distance of Arcturus	nearly	7,000,000

These distances are expressed in terms of the earth's distance from the sun as the measuring rod. If we wish to express the distance in miles we should multiply them by 93 millions. The distance of Arcturus is still uncertain, almost immeasurably great, and the same is true of all but about 100 of the stars.

It is common to express the distances of the stars and other heavenly bodies by the time it takes their light to travel to us. The speed of light is such as it would travel round the earth more than seven times in a second. If we could send a ray of light round by the Atlantic and Indian Oceans to Australia and back here over the Pacific, keep it going round and round, and make a little tap every time it passed us, these taps would follow each other so rapidly we could hardly move our fingers fast enough to make them.

Going at this speed a ray of light from the moon reaches us in about $1\frac{1}{4}$ seconds, and one from the sun in 8 min. 20 sec. It reaches Neptune from the sun in 4 h. 10 m.

The light from Alpha Centauri reaches us in $4\frac{1}{3}$ years; that from 61 Cygni in about 9 years. From most of the stars of average brightness that we see in the sky at night the time is between 100 and 200 years, often more. From the telescopic stars it ranges from a few hundred years to perhaps several thousand.

Real Speed of a Star. — When we know the distance of a star, its apparent proper motion, and its motion in the line of sight, we can determine the actual speed with which it is flying through space. Arcturus has the most rapid motion of any star visible to the naked eye. We can easily compute it in this way. Let ES be the radius of the earth's orbit and AR the distance through which Arcturus moves in a year. The

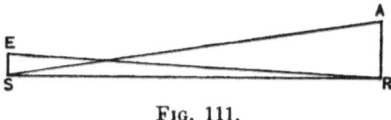

Fig. 111.

angle between the lines RE and RS is, as just explained, the parallax of the star, which, as we have already said, is about 0.03″. But the star moves through an arc AR, which is found to be 2.10″ per year. This is about 70 times as much as the parallax. It follows that the star Arcturus travels over more than 70 times the distance of the earth from the sun in a year, and therefore makes this distance in about 5 days. If we make the calculation we shall find that this corresponds to a speed of more than 200 miles per second!

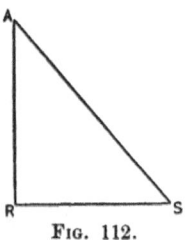

Fig. 112.

But this calculation takes no account of the motion of the star in the line of sight. To show the actual movement, we draw a line AR at right angles to the line of sight from the earth to the star, and make it proportional to the apparent motion as seen by the eye. We draw another line, RS, at right angles to this, showing the motion in the line of sight. Then, joining AS, the hypothenuse will represent the total real motion of the star.

In the case of Arcturus the motion in the line of sight is

so small that the hypothenuse differs very little from the line *AR*. We may therefore regard 200 miles per second as its probable speed.

Arcturus has undoubtedly been flying through space with this wonderful speed for thousands of years in the past, and will continue its journey for thousands of years in the future. We believe this to be the case because we cannot imagine or believe in the existence of any force which could change so rapid a motion of so immense a body.

6. Variable Stars. — The great majority of the fixed stars never change in their appearance. But there are some which vary in brightness from one time to another. These are called *variable stars*.

Some of these stars vary in so irregular a way that no law can be seen in their changes of light. The most remarkable case of this kind is that of Eta Argus, in the southern celestial hemisphere. Before 1830 it varied between the second and fourth magnitudes. From 1830 to 1843 it was sometimes nearly as bright as Sirius. Then it slowly faded away till it became invisible to the naked eye. In recent years it has brightened up a little, but in 1898 a telescope was still required to make it visible.

Most of the variable stars go through their changes according to a regular law, brightening up, and then fading away again in a regular period. These are called *periodic stars*.

The *phases* of a periodic star are the different appearances, or degrees of brightness, which it exhibits.

The *period* is the length of time in which it goes through its phases and returns to the same brightness as before.

A *minimum* is the phase when it gives less light than it did just before, or just afterward.

A *maximum* is the phase when, having been brightening up, it is about to fade again.

The law of variation is not the same in all periodic stars. The various kinds of change they go through are called *types*.

Thus, when we say that a star is of the *Algol type*, we mean that it varies in the same way that Algol does, a way that will be stated presently.

The period in the case of any one star commonly changes slowly from time to time, sometimes being a little longer, and sometimes a little shorter.

The periods of different stars are very different in length. In the case of some it is only a few hours, and in others a few days. In a great number of stars it ranges between six months and two years.

Interesting Variable Stars. — In the great majority of cases the variations in the light of these stars are so slight, or the stars are so faint, that only the most careful and well-trained observers would notice any change. But there are three of which the variations can be seen by any one who will take the necessary care in watching.

The Star Beta Lyræ. — One of these stars is in the constellation Lyra, and is marked with the Greek letter β (Beta) in the figure of that constellation (p. 203). One who notices the figure in the book can easily recognize the star in the heavens. If he looks at it for a few nights he will find that sometimes this star is of the same brightness as the star Gamma, close to it, and sometimes nearly a magnitude fainter.

It goes through a regular series of gradations, growing brighter and fainter at intervals of six days. The curious feature of this variation is that at every alternate minimum it is fainter than at the intermediate minima.

It is now believed that this star is composed of a pair of oval-shaped stars so very close together that they almost touch each other. The pair revolve around each other in an orbit of which the plane passes in the direction of the solar system. Hence, if we could have a telescope powerful enough to see what was going on, we should, at a certain time, see the small star pass in front of the bright one, partially obscuring it. Six days later the small one would pass behind the bright one and be hidden by it, and so on in regular order.

THE STARS AND NEBULÆ

The Star Algol. — Another remarkable variable star is Algol, or Beta Persei. This star is commonly between the second and third magnitude in brilliancy. But it fades away to nearly the fourth magnitude at intervals of about 2 d. 21 h. It takes 3 or 4 hours to thus fade away, and then in 3 or 4 hours more it brightens up again. It is now found that this is due to the star having a dark planet revolving around it, which is nearly as large as the star itself. Every time this planet passes in front of it it obscures part of its light. The fact of this revolution has been determined with the spectroscope by measuring the effect of the eclipsing planet upon the motion of the star. The planet itself is invisible even with the most powerful telescope.

Mira Ceti. — The third star of this kind is Omicron Ceti, called also *Mira Ceti*, or the wonder of Cetus. It is commonly invisible to the naked eye; but at intervals of about 11 months it gradually brightens up so as to be plainly visible. Sometimes it attains the second magnitude, sometimes only the fifth. But however bright it becomes, it begins to fade away again in two or three weeks and finally disappears from view. It may, however, always be seen with a telescope.

Stars of the Algol type are nearly always of the same brightness, but at certain regular intervals fade away for an hour or a few hours, and then brighten up again, as we have described in the case of Algol. There can be no doubt that these seeming eclipses arise from the same cause as those of Algol, namely, the presence of a dark planet which passes between us and the star at every revolution.

Sometimes two stars of the same kind revolve round each other. In this case, if the plane of the orbit passes through our solar system, we see them mutually eclipsing each other at every half revolution. That is to say, if we call the one star A and the other star B, then at one time of the revolution A will pass between us and B; half a revolution later A will pass on the other side of B, so that the latter will hide part of its light.

A remarkable pair of stars of this kind is called *Y Cygni*. Here the pair looks like one star, even in the most powerful telescope. The way we know there are two stars is that the eclipses occur at unequal intervals. The intervals are alternately 44 hours, 28 hours, 44 hours, and so on. This regular alternation shows that the two stars revolve round each other in an orbit having a great eccentricity, so that the revolution is much faster at one point of the orbit than at another.

The number of known stars of this type is very small. Commonly a variable star does not remain of the same brightness for any considerable time, but varies in a slow and regular way. Starting from the time when it is faintest, it grows brighter day after day and week after week, first at a very slow rate and then more rapidly, until it attains its full brightness, and then after a few days begins to fade away, slowly at first, and afterward more quickly, until it gets back to its least brightness, and so on in regular order.

Several hundred variable stars are now known to astronomers, and new discoveries of them are being made very rapidly. There is reason to believe that one star out of every hundred in the heavens may vary to a greater or less extent.

7. Double Stars. — A great many stars which seem single to the naked eye are found, when viewed with a telescope, to consist of two stars very close together. These are called *double stars*. Several thousand such stars are known, and new ones are constantly being discovered.

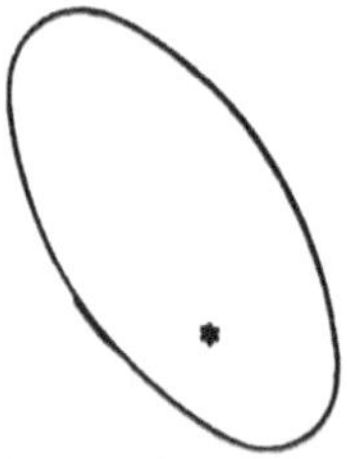

The first question suggested by a pair of this sort is whether the two stars are really close together, or whether they merely look so because they happen to lie in the same line from us. It is now known that nearly all such pairs are really close together and that the two stars of the pair revolve round each other. In this case the pair is called a *binary system*.

FIG. 113. — Orbit of a double star.

The time required for one star of a binary system to make a revolution round the other is called its *period*.

The period of most binary systems is very long, — several centuries, in fact. But a few have periods of less than a century. As accurate observations on these systems have only been made within a hundred years, only these few have been seen to complete a revolution. Hence the exact time of revolution is known only in those cases in which the period is less than a century, or not much greater.

Sometimes the two companion stars which form a double star, or binary system, are nearly of the same magnitude. In other cases, one star is very much smaller than the other. Indeed, a great many bright stars are found to have very minute satellites moving around them. Two of the most remarkable cases of this sort are Sirius and Procyon, because in each case the existence of the little satellite was inferred by the motion of the large star produced by the attraction of the satellite before the latter was seen by the telescope.

In the case of Sirius, it was found by Bessel and Peters that the visible star was moving round in such a way as to show that it was attracted by some body very close to it, which they could not see with their telescopes. But in 1862 Mr. Alvan Clark of Cambridge made a telescope of 18 inches diameter, which was more powerful than any ever before constructed. With this he found the companion of Sirius in the same direction in which it had been predicted from the motions of Sirius itself. It is now found that a revolution is made in about 50 years.

The history of Procyon is similar. Observations for a hundred years showed that it was moving in a little orbit, as if it were attracted by a dark body revolving round it in a period of 40 years. In 1896 this little body was actually found by Professor Schaeberle with the powerful telescope of the Lick Observatory, in California. It is found that the companion is moving round Procyon in the same direction in which the motion had been predicted.

A few of these binary stars will be seen double, even in quite a moderate telescope. But the greater number require a high telescopic power for their visible separation. The reason of this is that, in most cases, the stars are so close together that they appear single even with a high magnifying power, while in other cases the small one is so faint that it is obscured by the brightness of the larger one. With every increase of telescopic power, it is found that new objects of this class may be seen. The greater number of the difficult objects now known would have been entirely invisible in the largest telescopes of a hundred years ago. Hence we conclude that, if we could increase our telescopic power without limit, we should continually see more and more of these objects, and find perhaps that every star in the heavens had other stars revolving around it.

An interesting question now arises. Since our sun has a retinue of eight dark planets revolving around it, may it not be that all the stars have planets which we cannot see on account of their immense distance? This is quite possible. If we could fly away from the solar system, and carry with us the most powerful telescope ever made, all the planets, even the brightest, would become invisible through our telescope, one by one, long before we got halfway to the nearest star. Even Jupiter does not give $\frac{1}{1000000}$ part as much light as the sun, and we can readily understand that a star giving only $\frac{1}{1000000}$ as much light as another would be totally invisible to us. The fact that we cannot see planets revolving around the stars, therefore, proves nothing.

It might seem to us that it would be forever impossible that men living on this globe should be able to detect among the stars bodies which are forever invisible. And yet this wonderful thing is being done through the researches with the spectroscope and observations on the variable stars. We have already mentioned the variable stars of the Algol type, which are partially eclipsed by the revolution of dark bodies around them. Such stars will appear variable to us only in

THE STARS AND NEBULÆ

the rare cases when the plane of the orbit of the dark body passes near the direction of the solar system. If the plane should be in a different position, then the dark body, or the revolving star, would not seem to us to pass over the bright one at each revolution, but would pass above or below it, as the moon at conjunction with the sun may pass above or below it without eclipsing it. In these cases the bright star will not be a variable one, and our telescopes would give no indication that there was more than one body.

But the spectroscope now comes in and tells the story of companions which are and must forever be entirely invisible to human eyes. Systems made known in this way are called *spectroscopic binary systems,* or merely *spectroscopic binaries.*

There are two ways in which a star may be shown to form part of a binary system by means of the spectroscope.

One way consists in measuring the motion of the star in the line of sight. Sometimes this motion is found to vary in a regular period, increasing for a certain time, then decreasing, then increasing again, and so on. Sometimes it will move toward us for a certain interval, then nearly stop, then come toward us again. There can be but one possible cause for such a change of motion, the attraction of a body revolving around the star. In such a case each body would revolve around the common center of gravity of both, as we have described in the case of the sun and moon.

The other method rests on the same principle. We have said that the motion in the line of sight is shown by a slight displacement of the lines of the spectrum. Sometimes these lines are seen to be double for a period; then single; then double again in regular order. This arises from the fact that there are two stars moving around each other and showing the same lines. When one is moving toward us and the other from us a spectral line of one star is displaced in one direction and the corresponding line of the other is displaced in the opposite direction. Then two lines are seen instead of one. But when the stars move toward or from us with the same velocity only one line is seen, because the lines of the two stars are merged together.

The periods of these spectroscopic binary stars are generally only a few days, whereas, as we have already mentioned, those which we see double in the telescope have periods of many years. But we may suppose that binary systems having all intermediate periods really exist, though they cannot be seen double in the telescope. It is only a few years since these systems began to be discovered with the spectroscope, and we have not had time to detect those having a long period. Besides, such a system will not be seen even with a spectroscope, unless the invisible body which produces the motion has a great mass. If an astronomer on a distant star should observe our sun with a spectroscope he would never be able to detect any motion due to the attraction of the planets which revolve around it.

We may therefore conclude that planets are probably revolving around great numbers of stars; but up to the present time astronomers have not been able to learn much more about them than what we have just set forth. But it is wonderful that we should ever know anything at all about them.

8. Clusters and Nebulæ. Clusters of Stars. — We have already spoken of some of these clusters which can be seen by the naked eye, either as separate stars or as patches of milky light. Quite a number of them are seen in the Milky Way, especially the southern part, which is visible in the late summer or autumn.

Some of these clusters are among the most interesting telescopic objects. One of the most remarkable is the great cluster of Hercules. This is formed of thousands

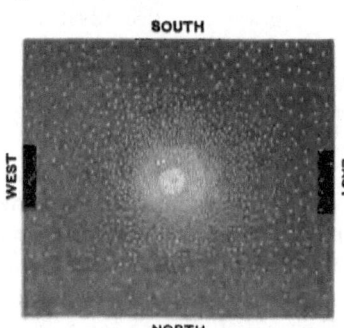

FIG. 114. — Star cluster 47 Toucani, as drawn by Sir John Herschel.

THE STARS AND NEBULÆ

of stars in such close proximity that they can scarcely be distinguished even in the most powerful telescopes. When we look at this object, we might imagine it to be a little colony on the outskirts of creation itself, the inhabitants of which might hold communication with each other, even if they lived on different planets. Yet, if any of the stars which form this cluster have planets revolving round them, it is quite likely that their distance apart is so great that the inhabitants of one planet would know no more about the other planets, or the other stars, then we know about Venus or Mars.

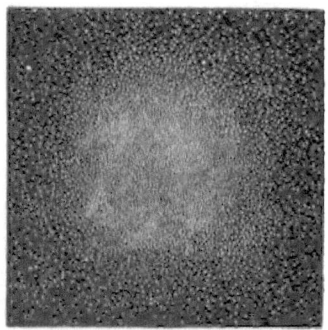

FIG. 115. — Star cluster ω Centauri, as drawn by Sir John Herschel.

Nebulæ. — There are in the heavens a great number of objects which appear in the telescope as very faint masses of soft diffused light, like thin pieces of cloud. Such a mass is called a *nebula* (Latin for cloud).

A few of these objects are visible to the naked eye, but the greater number, comprising many thousands, can be seen only with a good telescope and on very clear nights. Sometimes when a cluster of stars is viewed with the naked eye, or with a telescope which will not enable us to distinguish the separate stars, it has the appearance of a nebula. Hence it was once questioned whether all these objects might not really be clusters of stars. This question has been settled in recent times by the spectroscope, which shows that the greater number of the nebulæ are masses of glowing gas, and therefore cannot be made up of stars.

The most wonderful object of this sort is the great nebula of Orion. It is plainly visible to the naked eye, to which it has the appearance of a star of the fourth magnitude with a somewhat indistinct and hazy outline. It is the middle star of

222   *ASTRONOMY*

the three which form the sword of Orion, hanging below the belt, and may be found on the figure of that constellation already given. When examined in a good telescope a curious dark rift or cavity is seen near one edge. In this rift are a number of stars, of which four are much brighter than the others. These are called the *trapezium*. These stars give us the impression of being formed out of the matter of the nebula

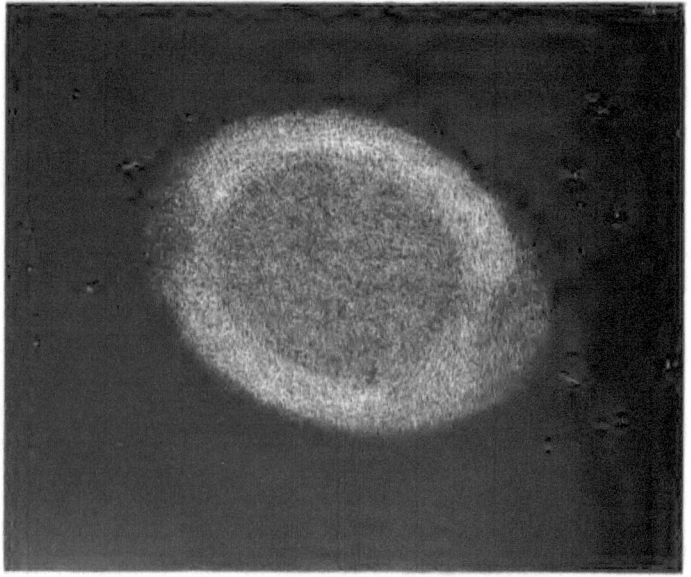

Fig. 116. — The annular nebula of Lyra.

which perhaps once filled the rift. There are a great number of other stars in and around the nebula, some of which are believed to be variable. Two very small stars are inside the trapezium.

Another of these objects easily visible to the naked eye is the great nebula of Andromeda. This does not look like a star, but can easily be seen as a small hazy patch of light of an elliptical form. It is sometimes mistaken for a small comet. With a small telescope it looks like a patch of fog illuminated by a lantern behind it or in it.

Many nebulæ are of singular or fantastic shapes. A celebrated one is the annular or ring nebula of Lyra, situated in that constellation about halfway between the stars Beta and Gamma. The circumference is so much brighter than the cen-

FIG. 117. — The Omega Nebula, as drawn by Sir John Herschel.

ter that, to the telescopes of former times, it seemed to be a true elliptical ring. But with our large telescopes the whole interior appears to be filled with nebulous light, as shown in the figure. The figures show several other curiously formed nebulæ, as drawn by Sir William Herschel and others. It is now found that there are many thousand nebulæ which are too faint to be seen even with a telescope, but which can be photographed

224   ASTRONOMY

by several hours' exposure. One of these is near the Pleiades. Another very large one is in the constellation Orion.

It is thought that stars and systems are formed by the gradual cooling and condensation of nebulæ. In this way the remarkable regularity of the solar system is explained. It is supposed that the matter composing the sun and planets was

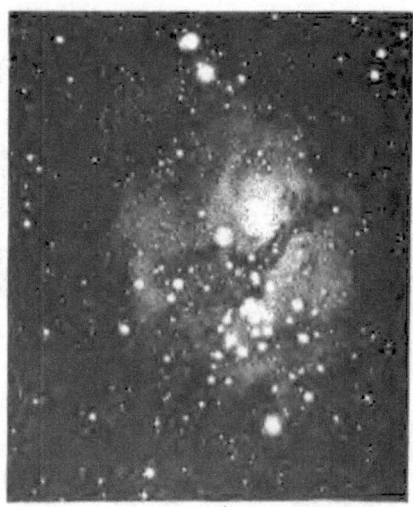

Fig. 118. — The Trifid Nebula.

once a mass of glowing gas, with a slow revolution on its axis. As this gas cooled, the outer portion condensed into a ring. As the central portion grew smaller, additional rings were formed. Finally, each ring contracted into a planet, and the central mass into the sun.

This view of the origin of the solar system is called the *nebular hypothesis*.

CHAPTER XVI

A BRIEF HISTORY OF ASTRONOMY

It is interesting to know what men thought of the earth and the heavens before their true relation and the laws of the celestial motions were understood.

In very ancient times it was well known to philosophers and navigators that the earth was a globe. Navigators could see how, on the Mediterranean Sea, the sail of a distant ship would gradually sink below the horizon, owing to the roundness of the surface of the sea. Observers of the heavens knew that when they traveled south the stars behind them would sink below the horizon, while new stars in front of them would rise. They also knew that an eclipse of the moon which was seen after sunset at Babylon occurred at an earlier hour at points farther west. All this proved to them that the earth on which we live is a globe.

What they did not know, and had no means of finding out, was that this globe turned on its axis and moved round the sun. It is sometimes supposed that Pythagoras and other ancient philosophers taught this system; but this cannot be proved, because none of their writings have come down to us. The earliest astronomer who has told us much of the ancient ideas of astronomy was Ptolemy, who flourished at Alexandria about the year 150 of our era. He wrote a great work called the *Syntaxis*, but more commonly known as *Almagest*, from an Arabic expression signifying *The Great Work*. The system of astronomy which we find in this book is commonly called the *Ptolemaic System*, after this writer. Its main doctrines are these: —

1. The earth is a globe.
2. This globe is at rest in the center of the heaven.
3. The heaven is spherical in form.
4. The earth is so much smaller than the heaven that it is only a point in comparison.
5. The heaven makes a revolution around the earth every day.

It is interesting to notice what is right and what is wrong in these five propositions. The first and fourth, as we already know, are quite correct; and the fact that the ancients were able to learn so much about the respective magnitudes of the earth and the heaven is very remarkable. The second is wrong. Properly speaking, there is no such thing as the center of the heaven. But the ancients, seeing the apparent circular motion of the stars, and noticing how they seemed to be spread out on the celestial sphere, supposed the heaven itself to be spherical.

The third and fifth propositions are, of course, all wrong.

To us it is much more reasonable to suppose that a comparatively small point like the earth is in motion and the much larger heaven at rest, than it is to suppose the reverse. The ancients thought it was the heaven and not the earth which moved, because they did not suppose the former to consist of the same kind of matter as the earth, and because they were not acquainted with the laws of motion. They supposed that there was an inherent tendency in all moving bodies to stop moving and come to rest. They reached this conclusion because that seemed to be the case with all bodies in motion around us. What they did not see was that this tendency to stop was not inherent in the motion itself, but arose from the fact that the motions of all bodies around us were constantly resisted by friction, the resistance of the air, and the contact with the earth.

Some ancient philosophers did really suggest that it was the earth which turned on its axis while the heaven remained at rest. But Ptolemy thought, that if the earth revolved on its axis, it must be turning with such speed that it would leave the air behind it, so that we should have a furious gale blowing from east to west. He could not conceive of earth, atmosphere, and everything on the earth running so smoothly together that we should be quite unconscious of any motion at all.

In watching the stars the ancient philosophers saw what to them was a curious fact, namely, that the stars preserved their relative positions to each other while they seemed to turn round the earth, just as if they were set in a hollow revolving sphere. So they imagined such a sphere, of which the substance was the purest and most translucent crystal, and which was therefore called the *crystalline sphere*. This sphere they fancied to turn on an axis passing through the center of the earth and coinciding with what we know to be the earth's axis. This was called the axis of the heaven. The sphere, in turning, was supposed to carry the stars with it.

But this sphere did not account for the motions of the planets. By the planets the ancients meant all the heavenly bodies, seven in number, which did not seem to revolve in the supposed crystalline sphere. These bodies, as we have already said, were the Sun, the Moon, Mercury, Venus, Mars, Jupiter, and Saturn.

Each of these planets was supposed to have its own sphere. They quite correctly supposed that the spheres of the planets must be inside the sphere of the fixed stars; but they had no idea how much farther the stars were than the planets.

They also noticed that the planets, excepting the sun and moon, moved in the celestial sphere, sometimes from west toward east and sometimes from east toward west. They accounted for this by supposing them to have two motions, one round the earth, and the other on an epicycle. The latter was a small circle the center of which was considered to move round the earth, while the planet moved around in the circle itself.

We now know that this apparent motion round the epicycle is really due to the motion of the earth round the sun, which makes the apparent motion of the planet sometimes retrograde and sometimes direct. But the ancients supposed it to be a real motion. This notion was held until the sixteenth century, when it was refuted by the great Copernicus.

Copernicus was born at Thorn, in Poland, in 1473. He studied at the University of Cracow and became a priest. He also acted for a short time as Professor of Mathematics in Rome. He thought for many years over the mystery of the heavenly motions, and saw clearly how easily they could be explained by supposing that the earth revolved round the sun and turned on its axis, instead of its being the sun and the heavens that moved. He spent many years in writing a great work in which this view was set forth, and all the calculations growing out of it were made. But he was so modest and so fearful of the prejudice that might be excited by a new system that he resisted all the entreaties of friends to publish his book, until he was about seventy years old. Then it was printed under the title *De Revolutionibus Orbium Cœlestium*. Copernicus died in the year 1543, on the very day that he received the first printed copy of his book.

FIG. 119. — Copernicus.

The system here set forth, which we now know to be the true one, is commonly called the *Copernican System*. Sometimes it is called the *Heliocentric System* because it makes the sun the center of motion; while the old Ptolemaic one is called the *Geocentric System* because the earth is the center of motion.

A BRIEF HISTORY OF ASTRONOMY 229

For a hundred years after the death of Copernicus his system was not generally accepted. The Church authorities feared that it was contrary to the Scriptures, and so were disposed to forbid its being taught as a truth, although they were willing astronomers should use it in their calculations, merely assuming it to be true. About 1620 Galileo was tried and imprisoned for publishing books in which the truth of the system was set forth. But this did not prevent intelligent men from believing in it.

About 1580 arose a celebrated astronomical observer, Tycho Brahe. He built a great observatory at a place which he called Uraniberg, on an island near Copenhagen. Through the patronage of the king of Denmark he was enabled to fit up his establishment with instruments of a size and precision before unknown. Unfortunately he did not fully accept the Copernican system of astronomy,

FIG. 120. — Kepler.

and in consequence his fame is not so great as it otherwise would be. What is still more unfortunate is that the telescope had not then been invented, and consequently his observations were not exact enough to be of use to his successors. Yet by their aid Kepler, who was a contemporary of Galileo, showed that the orbit of Mars round the sun was an ellipse, having the sun in one of its foci. Hence the laws of motion of the planets in ellipses, which we have already mentioned, are called Kepler's laws.

About 1680, as we have already said, Sir Isaac Newton showed that the motions of the heavenly bodies could be all explained by the theory of universal gravitation. English philosophers accepted this view immediately, but those on the continent of Europe were slow to follow. Descartes, another philosopher, claimed that the planets were carried around the sun, and the satellites round the planets, by their floating in an etheral medium which was kept in rotation like a whirlpool. Hence this view is called the theory of vortices. It was extensively held, but was at last abandoned when it became clear that the correct theory was that of gravitation. It now became a very interesting problem with mathematicians to demonstrate mathematically how the planets ought to move under the influence of the sun's attraction combined with their attraction on each other. One of the greatest men in this work was Laplace, who lived at Paris during the latter part of the eighteenth and the beginning of the nineteenth century. He published a great work called the *Mécanique Céleste*, in which the methods of solving the problem were set forth.

Fig. 121. — Sir Isaac Newton.

Let us again mention the four greatest books in which the laws of the celestial motions have been expounded: —

Ptolemy's *Syntaxis*, commonly called *Almagest;*
Copernicus *De Revolutionibus Orbium Cœlestium;*
Newton's *Principia;*
Laplace's *Mécanique Céleste*.

Observational Astronomy.

— We may now say something of the history of telescopes and observational astronomy. Before the time of Galileo, astronomers had to make all their observations with the naked eye, and with very crude and very inaccurate instruments. The first telescopes were made in Holland about the year 1608; but they were only very imperfect little spyglasses, and it does not seem that their makers ever thought of looking at the heavens with them.

FIG. 122. — Galileo.

The telescope is commonly thought to have been reinvented by Galileo, who heard that such an instrument had been made in Holland, and began to study how it must have been constructed. Whether he really invented it over again without help, in this way, or whether he saw a description of it is not certain. What is certain is that he was the first person to explain its principles, and to point it at the heavens and show what wonders it would make known to men. With his little instru-

FIG. 123. — Laplace.

232   *ASTRONOMY*

ments, poor as they were, he saw that the Milky Way was made up of countless stars. He also saw the phases of Venus, which proved that the planet was a dark globe which we see by the light of the sun. He saw the rings of Saturn projecting from the planet like two little handles, but could not see that they were rings. He saw the satellites of Jupiter, and found that they revolved round Jupiter as the planets did round the sun.

Huyghens, who flourished during the latter half of the seventeenth century, was the first to show the true form of the rings of Saturn.

The telescope has been gradually perfected and enlarged from the time of Galileo until the present. The greatest step forward was made by the celebrated Sir William Herschel, who observed in England during the latter part of the eighteenth century. . He made reflecting telescopes many times larger than any before made, and nearly as large as any made up to our time. With them he studied the starry heavens, and made greater discoveries than any one had done before him.

FIG. 124. — Sir William Herschel.

The first maker of an achromatic telescope was Dollond of London, who worked about 1760. But his glasses were only a few inches in diameter, because the art of making flint glass of good quality had not then been mastered. During the early part of the nineteenth century the most celebrated maker of refracting telescopes was Fraunhofer, of Germany. During its later part, from 1860 to 1890, the place of Fraunhofer was taken by Alvan Clark, of Cambridgeport, Massachusetts, and his two sons, Alvan and George. We have already spoken of the object glasses made by this family. Now, there are many able constructors of large telescopes in Europe and America.

ASTRONOMICAL WORK AT THE PRESENT TIME

The application of the spectroscope to astronomical observation, which commenced about the year 1860, has given rise to a new branch of astronomy known as *astrophysics*. This branch has been powerfully reënforced by the application of photography to astronomical observation. A telescope may be used like a very large camera. If we point it at the heavens and then place a sensitized plate in its focus, as the photographer puts such a plate in the focus of his camera, we may take a picture of any heavenly body at which the telescope is pointed. If we point the photographic telescope at the sky during the night, and make it follow the stars in their diurnal motion, we may thus take a picture of the stars in the field of view. The number of stars on the plate will depend on the length of the exposure. It is found that when the telescope is so set as to follow the diurnal motion of the stars very exactly, the image of the same stars may be kept on the same point of the plate during a period of several hours. In this case photographs will be found of many more stars than the eye can see with the same telescope. In other words, the power of the telescope is greatly increased by the extreme sensitiveness of the photographic plate.

We have already spoken of some remarkable nebulæ found in this way which would never have been known had we depended on the eye alone. Photography is now applied with success to spectroscopy by throwing the spectrum of a star upon a sensitized plate. A photograph can thus be made of a spectrum which the eye could scarcely see. This photograph can be studied by the astronomer at his leisure, and many conclusions deduced from it which he would not be able to deduce by a study of the spectrum itself with his eye.

It is possible to discover new objects in the heavens by photography with much greater facility than by the eye. We have already seen how minor planets are discovered by photography. No astronomer now attempts to find them in any other way. He simply points his telescope at any region of the heavens near the ecliptic, starts its clockwork so that it shall exactly follow the stars, and takes his photograph by an exposure of several hours. Then, if there is any planet in the field of view, it will not be photographed as a star, which is a mere point, but as a short line. This is because the planet has moved on the celestial sphere and left its trace upon the plate.

During an eclipse of the sun a few years ago, the impression of a little comet was found on the photographic plate, in the immediate neighborhood of the sun. What became of it is not known, because it was never seen by the eye. In one case a comet was actually discovered by photography and afterward observed by the eye.

In 1887 an arrangement was made among a number of observatories in the northern and southern hemispheres to make a complete photographic map of the heavens. To each observatory a certain region of the heavens was assigned. When this work is completed there will exist a set of several thousand photographs showing the starry heavens as they were about the end of the nineteenth century. How many millions of stars may be impressed on these plates we cannot yet say. Each region is photographed twice in order that there may be no doubt of the existence of anything that looks like a star on the photographic plate. Were only one plate taken, it is possible that sometimes a speck might be mistaken for a star. But if the same speck appears on two plates, then we know that a star must have been there. If then, at any future time, another photograph of the region is taken, it can be determined whether any star has disappeared or whether a new one has come into sight.

ASTRONOMICAL WORK AT THE PRESENT TIME 235

Quite separate from this international work is that of the Harvard Observatory at Cambridge, Massachusetts. Here a constant watch is kept on the heavens every clear night by a moving photographic telescope, which records automatically any new object as bright as the sixth magnitude that may come into view. Photographs of large regions of the heavens are also taken very often, so that changes in the brightness of the stars or planets before unknown may be detected.

Ever since the beginning of their science astronomers have been studying the motions of the heavenly bodies with a view of establishing the laws that govern them and the causes that may change them, and tracing the conclusions that may thus be drawn respecting the structure of the universe. We have told the main results in the preceding chapters of this book, but there are many important questions which we have not had time to discuss or explain. One is, whether the motions of the planets are influenced by any force except the gravitation of the sun and of the other planets. Mathematical methods have been brought to such perfection that the astronomer, when aided by exact observations, can predict the path which a planet ought to follow in consequence of the attraction of all known bodies, with great exactness, for many years ahead. Then, comparing his conclusions as to the place of the planet with observations, he can see whether his predictions are correct. If they are, he knows that the theory on which they are based is true. If the predicted positions do not agree with observation, he investigates the cause. We have seen what a brilliant result was thus obtained in the discovery of the planet Neptune.

There are some cases in which success has not yet been reached. We recall that the orbits of all the planets slowly change their positions in consequence of the attraction of each planet on all the others. It is found that the perihelion of Mercury changes its position somewhat more than it should in consequence of the attraction of the other planets. For some time it was supposed that this might be due to the attraction of unknown planets between Mercury and the sun. But it is now made almost certain that no planets of sufficient mass to produce the effect can exist. The cause of the motion is therefore still unknown.

It is also found that the motion of the moon changes very slowly and slightly from one century to another, sometimes going a little ahead, sometimes falling a little behind. The cause of these deviations has not yet been discovered.

Altogether, although so much has been learned about the heavenly bodies, there is still so much left to learn that the most advanced astronomer must feel as if he were only at the threshold of his science.

INDEX

Including references to definitions of the technical terms used in this book

Aberration, 60, 78, 79.
Absorption, selective, 74.
Achromatic, 61.
Aldebaran, the star, 199.
Almagest of Ptolemy, 225.
Altitude, 16.
Andromeda, the constellation, 195.
Angle of the vertical, 91.
Annual motion, 34.
Annual parallax, 209.
Annular eclipse, 129.
Annulus, 129.
Antarctic circle, 36.
Antipodes, 11.
Aphelion, 142.
Apogee, 120.
Apparent diameter, 74.
Apparent motion, 20.
Apparent noon, 51.
Apparent time, 51.
Apparition, circle of perpetual, 26.
Arctic circle, 36.
Aspects of a planet, 146.
Asteroids, 144.
Astronomical latitude, 91.
Astronomical refraction, 58.
Astronomical time, 139.
Astrophysics, 233.
Atmosphere, 100.
Atmospheric refraction, 58.
Auriga, the constellation, 195, 198.
Autumnal equinox, 37, 41.
Axis, 11.
Axis, declination, 63; polar, 63.

Base line, 94.
Bayer, system of naming stars, 196.
Betelguese, the star, 200.

Binary systems, 216, 219.
Body, 80.

Calendar, 133; Gregorian, 135; Julian, 135.
Canals on Mars, 157.
Cancer, tropic of, 36.
Canis Major, the constellation, 200.
Canis Minor, the constellation, 200.
Capella, the star, 198.
Capricorn, tropic of, 37.
Cassiopeia, the constellation, 195, 197.
Castor and Pollux, 200.
Celestial equator, 21.
Celestial horizon, 17.
Celestial poles, 21.
Celestial sphere, 12.
Central eclipse, 128.
Centrifugal force, 89.
Chromatic aberration, 60.
Circle, antarctic, 36; arctic, 36; meridian, 70.
Circle of perpetual apparition, 26.
Circle of perpetual occultation, 28.
Circumpolar, 197.
Civil time, 50, 139.
Clark, Alvan, 66, 233.
Clock, sidereal, 49.
Clusters of stars, 195.
Collimator, 72.
Coma of a comet, 176.
Comet, Biela's, 185; description of a, 176; Donati's, of 1858, 182; Encke's, 183; Halley's 180; great, of 1843, 182; great, of 1882, 183; periodic, 177.
Comets, constitution of, 185; orbits of, 178; remarkable, 180; telescopic, 177.
Conjunction, 117, 147.

INDEX

Constant of aberration, 79.
Constellation, 195.
Copernican System, 228.
Copernicus, work of, 228.
Corona of sun, 130.
Counter-glow, 102.
Crystalline sphere, 227.
Cycle, Metonic, 137; solar, 138.

Day, 11, 133; sidereal, 49.
Declination, 28.
Declination axis, 63; parallel of, 30.
Diameter, apparent, 74.
Direct motion, 148.
Dispersion, 58.
Distance, apparent, 13; linear, 13; mean, 141.
Diurnal motion, 20.
Dollond, 232.
Dominical letter, 137.
Double stars, 216.
Draco, constellation, 197.

Earth, magnitude of the, 99.
Easter Sunday, 134.
Eastern time, 54.
Eccentricity, 141.
Eclipse, partial, 125.
Eclipses of the moon, 124.
Eclipses of the sun, 126.
Ecliptic, 39; obliquity of the, 35; plane of, 35; pole of the, 46.
Ellipticity, 90.
Elongation of a planet, 146.
Equation of time, 51.
Equator, 11; celestial, 21; of the sun, 108; plane of the, 20.
Equatorial telescope, 62.
Equinox, autumnal, 37, 41; vernal, 36.
Equinoxes, precession of the, 45.
Equinoxial year, 44.
Eros, the planet, 161.
Eyepiece, 61.

Field of view, 62.
Focal length, 60.
Focus of an elipse, 141.
Force, 80; centrifugal, 89.
Fraunhofer, 66, 232.
Friction, 80.
Full moon, 119; Paschal, 134.

Galileo invents telescope, 231.
Gegenschein, 102.
Gemini, constellation, 200.
Geocentric latitude, 92.
Geocentric System, 228.
Geodesy, 93.
Georgium sidus, 172.
Geographical latitude, 91.
Geoid, 90.
Gibbous, 118.
Golden number, 137.
Gravitation, universal, 83.
Gravity, 10, 81.
Gregorian Calendar, 135.

Head of a comet, 176.
Heliocentric System, 228.
Hemisphere, invisible, 19; visible, 19.
History of Astronomy, 225.
Horizon, celestial, 15; dip of, 16; plane of, 15.
Horizontal parallax, 76.
Hour circle, 28, 30.
Hyades, 199.
Hypothesis, nebular, 224.

Image, 60.
Inertia, 82.
Inferior conjunction, 147.
Inferior planets, 144.
Invisible hemisphere, 19.

Julian Calendar, 135.
Jupiter, the planet, 162; satellites of, 165; surface of, 163.

Kepler, work of, 229.
Kepler's laws, 141.

Latitude, astronomical, 91; geocentric, 92; geographical, 91; parallel of, 30.
Letter, Dominical, 137.
Libration, 120.
Light, zodiacal, 101.
Line, of nodes, 126, 153; of sight, 62; of total eclipse, 127.
Line, vertical, 17.
Linear distance, 13.
Local time, 53.
Longitude, telegraphic, 97.
Lunar month, 134.

Major planets, 142.
Mars, canals of, 157; the planet, 156; rotation of, 159; satellites of, 159; supposed inhabitants of, 159.
Mass, 84.
Matter, 80.
Maximum of a variable star, 213.
Mean distance, 141.
Mean noon, 51.
Mean sun, 51.
Mean time, 51.
Mercury, the planet, 151; transits of, 153.
Meridian, 22.
Meridian circle, 70.
Meteoric showers, 188.
Meteoroids, 187.
Meteors, 187.
Metonic cycle, 137.
Mécanique Céleste, Laplace's, 230.
Mile, nautical, 93; statute, 93.
Minimum of a variable star, 213.
Minor planets, 144.
Moon, eclipse of the, 124.
Motion, annual, 34; apparent, 20; direct, 148; diurnal, 20; laws of, 81; proper, 192; relative, 10; retrograde, 148.
Mountain time, 54.
Mounting, 62.
Month, lunar, 134.

Nadir, 16.
Nautical mile, 93.
Neap tides, 123.
Nebula, 221.
Nebular hypothesis, 224.
Neptune, the planet, 173; satellite of, 174.
New moon, 118.
New style, 136.
Newton, Sir Isaac, 83, 230.
Nodes, 126; line of, 153.
Noon, apparent, 51; mean, 51.
Noon, sidereal, 49.
Nucleus, 106, 176.

Object glass, 61.
Objective, 61.
Obliquity of the ecliptic, 35.
Occultation, circle of perpetual, 28.

Olbers's hypothesis, 160.
Old style, 136.
Opposition, 147.
Orbits of the planets, 163.
Orion, the constellation, 200.

Parallax, 75, 76, 209.
Parallel, of declination, 30; of latitude, 30.
Partial eclipse, 125.
Pacific time, 54.
Paschal full moon, 134.
Pegasus, square of, 204.
Penumbra, in eclipse, 123; of sun spot, 106.
Perigee, 120.
Perihelion, 142.
Period, of a binary star, 217; of a variable star, 213.
Periodic comet, 177.
Periodic stars, 213.
Periodic time, 141, 144.
Perseids, 190.
Perturbations, 150.
Phases, of an eclipse, 129; of a variable star, 213.
Photography, celestial, 233.
Photosphere, 104.
Plane, of the ecliptic, 35; of the equator, 20.
Planet, primary, 144.
Planets, 33; 191; inferior, 144; major, 142; minor, 144; relative size of, 143; secondary, 144; superior, 144.
Pleiades, 199.
Polar axis, 63.
Poles, 11; celestial, 21; of the ecliptic, 46; of the sun, 107.
Polestar, 24.
Precession of the equinoxes, 45.
Primary planet, 144.
Principia of Newton, 230.
Procyon, the star, 200.
Prominences of the sun, 109.
Proper motion, 192.
Ptolemaic System, 225.

Radiant point, 188.
Radius vector, 141.
Reflecting telescope, 64.
Refraction, 56, 58.

INDEX

Relative motion, 10.
Retrograde motion, 148.
Revolution, sidereal, 117; synodic, 118.
Rigel, the star, 200.
Right ascension, 29.

Saturn, the planet, 167; views of, 167; satellites of, 170.
Secondary planets, 144.
Secular variations, 150.
Selective absorption, 74.
Semidiameter, 75.
Shadow, 123.
Shadow cone, 123.
Shooting stars, 187.
Showers, meteoric, 188.
Sidereal clock, 49.
Sidereal day, 49.
Sidereal noon, 49.
Sidereal revolution, 117.
Sidereal time, 49.
Sidereal year, 45.
Signs of the zodiac, 40.
Sirius, the star, 200.
Solar cycle, 138.
Solar spectrum, 71.
Solar system, 33.
Solar year, 44.
Solstice, summer, 41; winter, 41.
Spectroscope, 71.
Spectroscopic binary systems, 219.
Spectrum, 71; of a star, 71; solar, 71.
Spectrum analysis, 71.
Sphere, celestial, 12.
Spring tides, 122.
Standard time, 53.
Stars, shooting, 187; telescopic, 192.
Stationary, 148.
Statute mile, 93.
Style, new, 136; old, 136.
Sun distance, 145.
Sun, eclipses of the, 126; mean, 51; corona of, 129; density of, 102; equator of, 108; heat of, 105; mass of, 105; rotation of, 107; spots on, 106.
Superior conjunction, 147.

Superior planets, 144.
Synodic revolution, 118.
Syntaxis of Ptolemy, 225.

Tail of a comet, 176.
Taurus, the constellation, 198.
Telegraphic longitude, 97.
Telescope, equatorial, 62.
Telescopic comets, 177.
Telescopic stars, 192.
Tidal waves, 121.
Tides, 121–123.
Time, apparent, 51; astronomical, 139; central, 54; civil, 50, 139; eastern, 54; equation of, 51; local, 53; mean, 51; Pacific, 54; periodic, 141, 144; mountain, 54; sidereal, 49.
Total eclipse, line of, 127.
Transit instrument, 68.
Tropic, of Cancer, 36; of Capricorn, 37.
Tycho Brahe, observations of, 229.

Umbra, of sun spot, 106.
Uranus, the planet, 172; satellites of, 172.
Ursa major, the constellation, 195, 197.
Ursa minor, the constellation, 197.

Variable stars, 213.
Variations, secular, 150.
Venus, the planet, 153; rotation of, 155; transits of, 155.
Vernal equinox, 36.
Vertical, angle of the, 91
Vertical line, 17.
Vector, radius, 141.

Waves, tidal, 121.
Weight, 84.

Year, equinoxial, 44; sidereal, 45; solar, 44.

Zenith, 16; distance, 17.
Zodiac, 39; signs of, 40.
Zodiacal light, 101.

About the Author

Simon Newcomb (March 12, 1835 – July 11, 1909) was a Canadian-American astronomer and mathematician. Though he had little conventional schooling, he made important contributions to timekeeping as well as writing on economics and statistics and authoring a science fiction novel.

Newcomb studied mathematics and physics privately and supported himself with some school-teaching before becoming a functionary in charge of calculations at the Nautical Almanac Office in Cambridge, Massachusetts. In 1861, Newcomb took advantage of one of the ensuing vacancies to become professor of mathematics and astronomer at the United States Naval Observatory, Washington D.C. Newcomb set to work on the measurement of the position of the planets as an aid to navigation, becoming increasingly interested in theories of planetary motion.

By the time Newcomb visited Paris, France in 1870, he was already aware that the table of lunar positions calculated by Peter Andreas Hansen was in error. While in Paris, he realised that, in addition to the data from 1750 to 1838 that Hansen had used, there was further data stretching as far back as 1672. His visit allowed little serenity for analysis as he witnessed the defeat of French emperor Napoleon III in the Franco-Prussian War and the coup that ended the Second French Empire. Newcomb managed to escape from the city during the ensuing rioting that led up to the formation of the Paris Commune and which engulfed the Paris Observatory. Newcomb was able to use the "new" data to revise Hansens tables. He was offered the post of director of the Harvard College Observatory in 1875 but declined, having by now settled that his interests lay in mathematics rather than observation.

In 1877 he became director of the Nautical Almanac Office where, ably assisted by George William Hill, he embarked on a program of recalculation of all the major astronomical constants. Despite fulfilling a further demanding role as professor of mathematics and astronomy at Johns Hopkins University from 1884, he conceived with A. M. W. Downing a plan to resolve much international confusion on the subject. By the time he attended a standardisation conference in Paris, France in May 1896, the international consensus was that all ephemerides should be based on Newcombs calculations. A further conference as late as 1950 confirmed Newcombs constants as the international standard.

In 1878, Newcomb had started planning for a new and precise measurement of the speed of light that was needed to account for exact values of many astronomical constants. He had already started developing a refinement of the method of Léon Foucault when he received a letter from the young naval officer and physicist Albert Abraham Michelson who was also planning such a measurement. Thus began a long collaboration and friendship. In 1880, Michelson assisted at Newcombs initial measurement with instruments located at Fort Myer and the United States Naval Observatory, then situated on the Potomac River. However, Michelson had left to start his own project by the time of the second set of measurements between the observatory and the Washington Monument. Though Michelson published his first measurement in 1880, Newcombs measurement was substantially different. In 1883, Michelson revised his measurement to a value closer to Newcombs.

www.ingramcontent.com/pod-product-compliance
Lightning Source LLC
Chambersburg PA
CBHW022004160426
43197CB00007B/272